Glass Reinforced Plastics in Construction
Engineering Aspects

Glass Reinforced Plastics in Construction
Engineering Aspects

Leonard Hollaway

Department of Civil Engineering
University of Surrey

Surrey University Press
in association with International Textbook Company

Published by Surrey University Press
Bishopbriggs, Glasgow G64 2NZ

First published 1978

ISBN 0 903384 21 3

Printed in Great Britain by Robert MacLehose & Co. Ltd.
Printers to the University of Glasgow

To Patricia and Suzanne

Foreword

This book forms part of a series covering various aspects of design and materials for the engineer.

The emphasis is on the word *design* as this is the crucial part of an engineer's activity, which distinguishes him from a scientist working in the related field. The engineer must be able to synthesize the information available for the purpose of creating on paper that which is to be built, i.e. designing. This activity requires knowledge of analysis of the whole as well as of the components of that which is being designed. It requires also knowledge of the codes and standards in force and what is considered good engineering practice. Finally, design requires engineering judgement, and judgement can come only with experience.

This is why a university course, even if design-oriented as at a few universities, does not produce a designer. The young graduate, however bright, when confronted with his first, and possibly not only first, design job simply lacks the background for his new task. This is where the books in the present series come in: they make it possible to approach the task of design in a reasonable manner.

The books cover mainly the field of structural design and include works on precast concrete, prestressed concrete and composite materials and formwork. There lie the design problems that the civil engineer is concerned with. The books explain relatively simply the background to the design problem, and then in quite some detail the main features of the design process. All this is fully illustrated by worked examples. Such an approach may seem old-fashioned to an enthusiast of pure analysis but example and precept are essential if modern design is to build upon the accumulated stores of successful design used in the past. Indeed, modern disciplines have borrowed our approach but re-named it 'case study'.

The books in the present series should thus prove of great value to the young engineer and also to his slightly senior colleague who is designing in an unfamiliar field. For these people the book is a 'must'. But it is also a wise investment (and in these inflationary days such is not easy to come by) for the undergraduate who appreciates the importance of design. With the aid of the book he can profit more from his university or polytechnic course and enter employment better prepared.

The authors of the books in the series are all specialists willing, in the best spirit of the engineering profession, to share their knowledge and experience.

I am therefore confident that the series will be a success.

I can make these complimentary remarks as I have served only as general editor and not as author. There is therefore little doubt that the present series not only will fill a genuine need but will do so really well.

A. M. Neville
Dundee, 1978

Contents

Preface

This book is intended for engineers and architects who need to acquaint themselves with the properties, advantages, limitations and peculiarities of glass reinforced polyesters.

Many fabricators of this material and suppliers of the raw materials are not familiar with the needs of the civil and structural engineer and therefore are not able to help him as effectively as required. There is no convenient handbook that puts the necessary information in one place and it is hoped that this book will, to some extent, fill the gap in the literature.

The aim of the book is to introduce this composite material and to provide a simple guide to the principal aspects of the theory and use of G.R.P. in the construction industry. There is a vast amount of information on the topic but it tends to be scattered and uncoordinated. Some papers on specific topics have been discussed in the book, although it is not claimed that they present an exhaustive picture; however it is hoped that they will provide a starting point for any further investigations of the literature required.

Chapter 2 gives an introduction to fibre-matrix composites and their basic characteristics. Chapter 3 serves as an introduction to the components of G.R.P. and the various fabrication techniques; the terms used are those with which engineers and architects are familiar. They are not treated in any way from a chemical aspect.

The mechanical properties of structural plastics and G.R.P. composites are introduced in Chapter 4. As composites are manufactured from two constituents, the properties of these will be reflected in the overall properties of the composite. Therefore, in this chapter the common methods of testing plastics materials are fully discussed because the relevance of the test procedures must be understood before the behaviour of the material is appreciated.

Chapter 5 is an introduction to the in-service properties of G.R.P. composites. Investigations into the strength and durability of plastics under continuous stress have commenced at the Building Research Establishment and the results for the past two years are discussed in this Chapter. Also in Chapter 5, Part III is devoted to the fire tests applicable to G.R.P. as laid down in BS 476 and the behaviour of the material is discussed. The engineer and architect are not generally directly involved in the fire test but it is essential for them to understand the procedure and the relevance of the results.

Chapter 6 introduces the sandwich construction consisting of two faces of G.R.P. and a foam core.

Chapter 7 discusses architectural features used in G.R.P. structural configurations.

Chapters 8 and 9 give various methods and worked examples which may be used to analyze the complete G.R.P. structure. Because the overall structure tends to be of the folded plate configuration of complex form, the analysis is complicated; the analysis of the equivalent skeletal structure will give only approximate results whereas the finite element analysis will be expensive to perform. Probably a large percentage of the design work performed by engineers is based upon intuition and an understanding of the performance of the structure under load; if this design sense can be developed for G.R.P. there is no reason why engineers should have reservations about designing in the material.

The information on which this book is based has been derived from many sources, including individuals, firms, and organizations and I have attempted to acknowledge their assistance in appropriate parts of the text. It would be quite impossible for me to express here my gratitude to everybody concerned for the assistance they have given but I am particularly indebted to the following: Dr. H. Allen, University of Southampton, Southampton; Dr. J. Crowder, Building Research Establishment, Garston; Mr. Richard Hodge and Mr. T. Shears, Peter Hodge and Associates, Crewkerne; Mr. A. Leggatt, Nachshen Crofts and Leggatt Consulting and Structural Engineers, London; Mr. R. Ogorkiewicz, Imperial College, London; Mr. L. Phillips, Royal Aircraft Establishment, Farnborough; Mrs. B. Rogowski, Fire Research Station, Borehamwood; Mr. N. Southam, Gollins Melvin Ward Partnership, London; Mr. H. Zonsveld, formerly of Shell Chemical Co. Ltd., London.

I would also like to extend my thanks for their help to my colleagues at the University of Surrey: Dr. U.K. Bunni; Mr. J. Kolosowski and Professor Z.S. Makowski.

University of Surrey
January 1978

1 Introduction

The strength properties of plastics materials are reasonably good in civil engineering terms but their stiffness, when measured by their modulus of elasticity, is low. For structural applications, in which both the strength and stiffness of the material are critical, it is therefore necessary to combine plastics with other materials into composites whose properties transcend those of the constituents. The most commonly employed component is in a particulate or a fibrous form.

In the particulate composites, particles of a specific material or materials are embedded in and bonded together by a continuous matrix of low modulus of elasticity. In fibrous composites, fibres with high strength and high modulus are embedded in and bonded together by the low modulus continuous matrix. The fibrous reinforcement may be orientated in such a way as to provide the greatest strength and stiffness in the direction in which it is needed, and the most efficient structural forms may be selected by the mouldability of the material. To increase the stiffness of the composite still further, the structural units which make up the complete structure may be folded so that the stiffness of the structure is derived from its shape as well as from the material itself.

In the construction industry, glass fibre and polyester resin are used to form the fibrous composite which is then known as glass reinforced polyester (G.R.P.). In the past this abbreviation has referred to glass reinforced plastics in general, but now it refers specifically to polyesters. This material has been developed since the second world war with the main growth, interest and technology beginning in the 1960's. The production of resins, catalysts and accelerators which cure at room temperature has facilitated the manufacture of G.R.P. by relatively straightforward techniques, using the open mould process without the need to provide presses and steel moulds.

G.R.P. is now well established in the construction industry and takes its place alongside materials such as metal and timber; it cannot be considered a cheap substitute for other materials.

Two sophisticated G.R.P. structures have played a major role in the

Figure 1.1 Dome structure in Benghazi

Figure 1.2 Roof structure of Dubai Airport

development of this material; these are the dome structure erected in 1968 in Benghazi (Figure 1.1) and the roof structure to the Dubai Airport built in 1972 (Figure 1.2). The design and fabrication for the latter structure took place in the U.K. and the units were then shipped to Dubai.

One of the reasons why engineers and architects have so far been reluctant to use G.R.P. stems from the adverse publicity plastics have received regarding their resistance to fire. In this context it is important to realize that there are many different types of plastics, all with specific mechanical and physical properties, and that a very limited number of these are used as structural materials in the civil engineering industry. G.R.P. is the main material and this may be used as a load bearing composite, with or without incorporating additives to give fire resistance properties complying with national and international standard specifications.

Another reason is that the fabricators of G.R.P. load bearing and infill units are to some extent unfamiliar with the construction industry and therefore they tend to lack the experience necessary for estimating the erection costs and the extras incidental to most construction sites. Furthermore, it is uneconomic to produce individual units, continuous production lines being much more profitable.

However, considerable advances have recently been made in the use of this material for the building industry. In addition to the all-plastics structures, incorporating features which, at first sight seem to be futuristic in terms of elegance and apparent extravagance, but which in reality are usually of vital structural importance, G.R.P. is increasingly being used for less esoteric applications, such as pipe work, drainage, ducting, etc. The establishment of G.R.P. for use as load bearing and infill panels has also increased.

A recent survey [1.1] has confirmed a growing awareness that attention should be paid to the fabrication of G.R.P., with emphasis on continuous production, if the construction industry is to take full advantage of the versatility of the material. In assessing the viability of using the material account must be taken of the following:

(a) the necessity for a substantial capital investment;
(b) the need to limit the size of components;
(c) the repetition of units as related to tooling costs;
(d) the form of construction; i.e. single skin or sandwich;
(e) the inclusion of other materials for various reasons;
(f) the fire resistance of the units.

Clearly, it is possible to use semi-mechanical processes to produce units which meet the technical and commercial requirements provided that the repetition of the component is assured. G.R.P. load bearing and infill panels are probably the major outlet for this material; however, the majority are manufactured by the hand lay method which is labour intensive. If these panels were manufactured in sufficient numbers by mechanical processes, the economics could be very favourable.

The following chapters deal specifically with various aspects of the analysis and design of G.R.P. as used in the construction industry, and should enable the engineer and architect to understand the basic principles behind its use. This in turn will, it is hoped, encourage them to use this architecturally exciting composite material or at least consider its use for specific applications.

References

1.1. JAAFARI, A., HOLLAWAY, L. & BURSTALL, M., 'Analysis of the use of glass reinforced plastics in the construction industry', The Reinforced Plastics Congress 1976, Innovation—The Basis of Reinforced Plastics Progress, The British Plastics Federation, Brighton, November, 1976.

2 Mechanism of fibre reinforcement and material design

2.1 Introduction

Composites of strong stiff fibres in a polymer resin, used as engineering materials, require scientific understanding in order that design procedures may be developed. The mechanical and physical properties of the composites are clearly controlled by their constituent properties and by the micro-structural configurations. It is therefore necessary to be able to predict properties under varying conditions. The most important aspect of composite material design is anisotropic behaviour, and it is necessary to give special attention to the methods of controlling this property and its effect on analytical and design procedures.

Research work undertaken in recent years has resulted in a clearer understanding of the characteristics of fibre/matrix composites. It has been demonstrated that, with the correct process control and a soundly based material design approach, it is possible to produce composites which can satisfy stringent structural requirements. However, because the durability of the composite over a period of say 50 years cannot be scientifically estimated, great care must be exercised in predicting stress and deformations over this time span.

2.2 Mechanism of reinforcement in fibre reinforced plastics

The reinforcement of a low modulus matrix with high strength, high modulus fibres uses the plastic flow of the matrix under stress to transfer the load to the fibre; this results in a high strength, high modulus composite. The aim of the combination is to produce a two phase material in which the primary phase which determines stiffness and is in the form of particles of high aspect ratio (i.e. fibres) is well dispersed and bonded by a weak secondary phase (i.e. matrix). The principal constituents which influence the strength and stiffness

4

of the composites are the reinforcing fibres, the matrix and the interface. Each of these individual phases has to perform certain essential functional requirements based on their mechanical properties so that a system containing them may perform satisfactorily as a composite.

2.2.1 Fibres

The desirable functional requirements of the fibres in a fibre reinforced composite are:

(a) the fibre should have a high modulus of elasticity in order to give efficient reinforcement;
(b) the fibre should have a high ultimate strength;
(c) the variation of strength between individual fibres should be low;
(d) the fibres should be stable and retain their strength during handling and fabrication;
(e) the diameter and surface of the fibres should be uniform.

2.2.2 Matrix

The matrix is required to fulfil the following functions:

(a) to bind together the fibres and protect their surfaces from damage during handling, fabrication and during the service life of the composite;
(b) to disperse the fibres and separate them so as to avoid any catastrophic propagation of cracks and subsequent failure of the composite;
(c) to transfer stresses to the fibres efficiently by adhesion and/or friction (when composite is under load);
(d) to be chemically compatible with fibres over a long period;
(e) to be thermally compatible with fibres.

2.2.3 Interface

The interface between the fibre and the matrix is an anisotropic transition region exhibiting a gradation of properties. It must provide adequate chemically and physically stable bonding between the fibres and the matrix. Its functional requirements vary considerably according to the performance requirement of the composite during its various stages under service conditions.

2.3 The relationship between the structure of the component and its property

In analyzing fibre reinforced matrix materials the primary aim is to obtain predictions of the average behaviour of the composite from the properties of the components. Of particular interest to the engineer are the mechanical

B

properties describing strength and stiffness. The factors which influence these characteristics are:

(a) the mechanical properties of the fibre and matrix;
(b) the fibre volume fraction of the composite;
(c) the degree of fibre matrix interfacial adhesion;
(d) the fibre cross section;
(e) the fibre orientation within the matrix.

A brief review will now be given of some of the fundamental considerations regarding the mechanism of reinforcement for both continuous and discontinuous fibre reinforced composites.

2.4 Composites consisting of continuous fibres

2.4.1 Stiffness characteristics

The fibres are dispersed throughout the composite and it will be assumed that they are uniform, continuous and unidirectional and that there is complete bond at the interface between the fibre and matrix. Such a basic composite is shown in Figure 2.1, and it is assumed that the orthotropic layer has three mutually perpendicular planes of property symmetry; it is characterized elastically by four independent elastic constants. They are:

E_{11} = modulus of elasticity along fibre direction
E_{22} = modulus of elasticity in the transverse direction
v_{12} = Poisson's ratio (transverse strain caused by longitudinal stress)
G_{12} = longitudinal shear modulus
v_{21} is obtained from the equation $v_{21} E_{11} = v_{12} E_{22}$

2.4.1.1 Longitudinal stiffness
Longitudinal stiffness will be introduced by considering the way in which an applied composite force P_c is distributed between the fibre and matrix and by studying the relationship between the two components and the elastic constants of the orthotropic composite.

The composite load P_c is shared between the fibre load P_f and matrix load P_m in the following way:

$$P_c = P_m + P_f \tag{2.1}$$

or $\quad \sigma_c A_c = \sigma_m A_m + \sigma_f A_f \tag{2.2}$

$$\sigma_c = \sigma_m V_m + \sigma_f V_f \tag{2.3}$$

where A is the area of the phase
V is the volume fraction with $V_c = 1$
σ is the stress in the phase
suffices m, f, and c refer to the matrix, fibre and composite respectively.

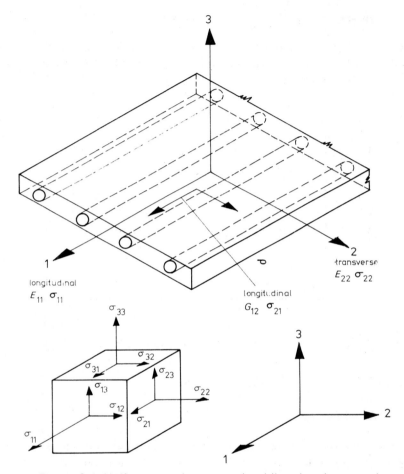

Figure 2.1 Uniform, continuous and unidirectional composite

As there is full bond between the phases, if ϵ is the strain in a phase:

$$\epsilon_c = \epsilon_m = \epsilon_f \tag{2.4}$$

Equation (2.3) may be written

$$\sigma_c = E_m\, \epsilon_c\, V_m + E_f\, \epsilon_c\, V_f \tag{2.5}$$

or $\quad \sigma_c = E_m\, \epsilon_c\, V_m + E_f\, \epsilon_c\, (1 - V_m) \tag{2.6}$

Since $V_f + V_m = 1$, the ratio of the load carried by the reinforcement to the load carried by the matrix is

$$\frac{E_f \, \epsilon_c \, (1 - V_m)}{E_m \, \epsilon_c \, V_m} = \frac{E_f}{E_m} \, \frac{(1 - V_m)}{V_m} = \frac{E_f}{E_m} \, \frac{V_f}{(1 - V_f)} \tag{2.7}$$

Figure 2.2 shows a plot of ratio of reinforcement load to matrix load as a function of the ratio between fibre and matrix modulus and as a function of fibre volume per cent.

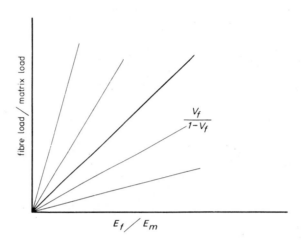

Figure 2.2 Relationship between (fibre load)/(matrix load), E_f/E_m and $V_f/(1 - V_f)$

From equation (2.6) it may be seen that

$$E_c = E_{11} = E_m \, V_m + E_f \, (1 - V_m) = E_m \, (1 - V_f) + E_f \, V_f \tag{2.8}$$

This equation is the well known 'law of mixtures' and has been found experimentally to predict E_{11} with a fair degree of accuracy.

2.4.1.2 Transverse stiffness

The modulus of elasticity E_{22} may be found using various analytical approaches. Shaffer[2.1] has derived the equation:

$$E_{22} = \frac{E_m}{1 - V_f(1 - E_f/E_m)} \tag{2.9}$$

which predicts E_{22} with reasonable accuracy.

2.4.1.3 Longitudinal shear modulus

The shear modulus value for unidirectional fibres has been investigated by several authors and Hashin and Rosen [2.2] have derived the following expression for circular cross section fibres arranged in an hexagonal array:

$$\frac{G_{12}}{G_m} = \frac{(G_f + G_m) + (G_f - G_m)V_f}{(G_f + G_m) - (G_f - G_m)V_f} \tag{2.10}$$

where G_f, G_m, G_{12} refer to the shear modulus of fibre, matrix and composite respectively.

If $G_f \gg G_m$ the equation may be simplified to

$$G_{12} = G_m \frac{(1 + V_f)}{(1 - V_f)} \tag{2.11}$$

A more rigorous analysis has been undertaken by Sendeckyj [2.3] in which he considers intrinsic composite variables such as random distribution, variation in character of fibres and the range of fibre modulus.

2.4.2 Strength characteristics

2.4.2.1 Uniaxial tension

From the previous section it will be seen that to obtain high stresses in fibre reinforced matrix materials and therefore to use high strength reinforcement most efficiently, it is necessary for the fibre modulus to be much greater than the matrix modulus. In addition, the volume fraction of fibres in the composite is in direct proportion to the composite load.

The excellent strengths and strength to weight ratios achieved by fibre reinforced matrix materials are a result of the high strength of the fibres, which are imparted to the composite. It is worth mentioning that the modular ratio (E_f/E_m) for glass fibre reinforced plastics is approximately 20 and at only 10% by volume of glass fibre, the glass assumes 70% of the total load.

Resin matrices are essentially brittle materials and have a higher ultimate fracture strain σ_{mu} compared to the ultimate failure strain of the glass fibre. Therefore at the ultimate strength of the composite the matrix is uncracked and carries its portion of the load as well as redistributing the loads by shear and large deformations from the broken fibres to their adjacent unbroken ones (see section 2.8.2). Assuming the matrix behaves elastically and that the composite fracture occurs at a strain corresponding to the fibre fracture strain the average stress in the composite at fracture is:

$$\sigma_c = \sigma_f V_f + \sigma_m (1 - V_f) \tag{2.12}$$

In the above equation the value of V_f must be greater than a certain minimum value. For small values of V_f the behaviour of the composite may not follow equation (2.12). This is because there is an insufficient number of fibres to effectively restrain the elongation of the matrix so that the fibres are rapidly stressed to their fracture point. If it is assumed that the fibres all break at the same stress, the composite fails unless the remaining matrix $(1 - V_f)$ can support the full composite stress. Therefore, failure of all the fibres results in immediate failure of the composite only if

$$\sigma_c = \sigma_f \, V_f + \sigma_m \,(1 - V_f) > \sigma_{mu} \,(1 - V_f) \tag{2.13}$$

This equation defines a minimum fibre volume fraction, which must be exceeded if the strength of the composite is to be given by equation (2.12).

The law of mixture relationship, equation (2.12), is an over-simplification and, although in some cases it is a good approximation, it takes no account of the statistical scatter in fibre strengths and the influence of the micro-structural defects. The effect of these quantities may be included in the equation (2.12) if it is modified to:

$$\sigma_c = \beta \sigma_\beta \, V_f + \sigma_m \,(1 - V_f) \tag{2.14}$$

where σ_β = fibre bundle strength

β = is the matrix efficiency factor, usually between 1 and 2

σ_m = the matrix stress at fibre fracture.

Consequently, it may be seen that the composite strength is principally dependent on fibre strength and volume fraction since the contribution from the matrix is small.

2.4.2.2 Uniaxial compressive strength

The mechanism of failure of an unidirectional fibre composite subjected to a compressive load is a difficult quantity to estimate and highly complex to analyze. Most of the effort to solve the problem has been directed towards the collection and semi-empirical representation of data for use in design. Dow [2.4], Rosen [2.5] and Foye [2.6] have all attempted to analyze the fracture mode of uniaxially aligned continuous fibre reinforced systems.

The problem may be treated similarly to that of the buckling of slender columns on an elastic foundation; the critical composite compressive stresses are given by:

$$\sigma_c = \frac{G_m}{1 - V_f} \qquad \text{shear mode, in phase buckling} \tag{2.15}$$

$$\sigma_c = V_f \left[\frac{2 \, E_f E_m \, V_f}{3(1 - V_f)} \right] \begin{array}{l} \text{extension mode, out of phase} \\ \text{buckling} \end{array} \tag{2.16}$$

In the practical range of V_f the shear mode tends to dominate and the compressive strength is highly dependent upon the properties of the matrix.

2.4.2.3 Transverse tensile strength

As may be expected the tensile strengths of fibre composites in a perpendicular direction to the fibres depends largely upon the tensile strength of the fibre matrix interface and the tensile strength of the matrix. In order to estimate the ultimate transverse tensile strength of a composite, Cooper and Kelly [2.7] have used the simple expression

$$\sigma_{transverse} = \sigma_m \left[1 - \sqrt{\left(\frac{4V_f}{\pi}\right)} + \sigma_1' \sqrt{\left(\frac{4V_f}{\pi}\right)} \right]$$
(2.17)

where σ_1' is the tensile stress required to rupture the fibre matrix load.

2.5 Discontinuous fibres

The behaviour of composites reinforced with fibres of finite length l cannot be described by equations 2.12 and 2.13 unless the length l is much greater than a critical length l_c (see Appendix, p. 219); thus in the above equations the value of σ_f or V_f requires adjustment.

When a composite containing uniaxially aligned discontinuous fibres is stressed in tension parallel to the fibre direction (Figure 2.3), there is a portion

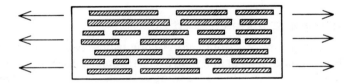

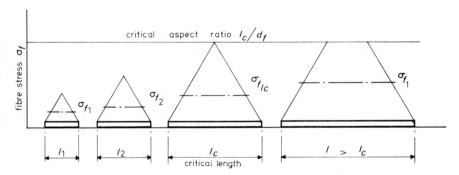

Figure 2.3 Schematic representation of discontinuous fibre/matrix composite. Average fibre tensile stress $\sigma_{f\,av}$ (σ_{f1} etc.) along the fibre length

at the end of each finite length which is stressed to less than the maximum stress of a continuous fibre. The critical transfer length over which the fibre stress is decreased from $\sigma_{f\,max}$ to zero is referred to as $\frac{1}{2}l_c$ and a quantity α may be defined as the ratio of the area under the stress distribution curve over the length $\frac{1}{2}l_c$ to the area of the rectangle represented by the product:

$$\sigma_{f\,max} \times \tfrac{1}{2} l_c$$

The fibre end of length $\tfrac{1}{2} l_c$ may be considered as supporting a reduced average stress $\alpha \sigma_{fmax}$ or may be reduced to an effective length of $\tfrac{1}{2}\alpha l_c$ subjected to a stress σ_{fmax}.

The average stress in the discontinuous fibre is less than either the ultimate strength, or the maximum stress, depending upon the fibre length.

The average stress over length l is given by:

$$\sigma_{f\,av} = \frac{1}{l} \int_0^l \sigma_f dl \quad \text{where } \sigma_f = \text{stress in fibre} \tag{2.18}$$

Integrating equation (2.18) will yield the average stress in the fibre strained to its ultimate failure; this stress is given by

$$\bar{\sigma}_{fav} = \sigma_f \left[1 - (1-\alpha)\frac{l_c}{l} \right] \tag{2.19}$$

2.6 Composite analysis

During the manufacture of a fibre/matrix composite it is usual to introduce multidirectional reinforcement and in so doing to increase the thickness of the composite to the required value; depending upon the structural requirement of the composite, the reinforcement could be any of the varieties described in the next chapter. It is desirable, therefore, when designing a composite to analyze the individual lamina characteristics and to apply the appropriate failure criteria to the individual lamina stresses in the principal directions; these stresses being transformed from the applied composite forces.

2.6.1 Homogeneous orthotropic materials

The stress-strain relationship for a unidirectional and bidirectional composite, may be established by assuming the laminae to be homogeneous orthotropic materials in a state of plane stress. The assumption in the macromechanics approach is that the fibre/matrix geometry and interactions are ignored.

The Hooke's law relationships are:

$$\sigma_{11} = Q_{11} \epsilon_{11} + Q_{12} \epsilon_{22} \tag{2.20}$$

$$\sigma_{22} = Q_{12} \epsilon_{11} + Q_{22} \epsilon_{22} \tag{2.21}$$

$$\sigma_{12} = Q_{66} \epsilon_{12} \tag{2.22}$$

where σ_{11} and σ_{22} are the normal stresses and σ_{12} is the shear stress, ϵ_{11} and ϵ_{22} are the normal strains and ϵ_{12} is the shear strain.

Or in the matrix form:

$$
\begin{bmatrix} \sigma_{11} \\ \sigma_{22} \\ \sigma_{12} \end{bmatrix} = \begin{bmatrix} Q_{11} & Q_{12} & 0 \\ Q_{12} & Q_{22} & 0 \\ 0 & 0 & 2Q_{66} \end{bmatrix} \begin{bmatrix} \epsilon_{11} \\ \epsilon_{22} \\ \tfrac{1}{2}\epsilon_{12} \end{bmatrix} \qquad (2.23)
$$

where Q is the stiffness matrix with components

$$
Q_{11} = \frac{E_{11}}{1 - v_{12}\,v_{21}}
$$

$$
Q_{22} = \frac{E_{22}}{1 - v_{12}\,v_{21}}
$$

$$
Q_{12} = \frac{v_{12}\,E_{22}}{1 - v_{12}\,v_{21}}
$$

$$
Q_{66} = G_{12}
$$

where E_{11}, E_{22}, G_{12} and v_{12} are the elastic constants of the composite.

The above implies that only four elastic constants are required to characterize an orthotropic material under plane stress.

If the lamina principal axes (1, 2) do not coincide with the reference axes (x, y) then the above constitutive relationships for each individual lamina must be transformed to the reference axes. This may be achieved by applying the following transformation:

$$
\begin{bmatrix} \sigma_{11} \\ \sigma_{22} \\ \sigma_{12} \end{bmatrix} = [T] \begin{bmatrix} \sigma_{xx} \\ \sigma_{yy} \\ \sigma_{xy} \end{bmatrix} \qquad (2.24)
$$

$$
\begin{bmatrix} \epsilon_{11} \\ \epsilon_{22} \\ \tfrac{1}{2}\epsilon_{12} \end{bmatrix} = [T] \begin{bmatrix} \epsilon_{xx} \\ \epsilon_{yy} \\ \tfrac{1}{2}\epsilon_{xy} \end{bmatrix} \qquad (2.25)
$$

The transformation matrix T is given by

$$
[T] = \begin{bmatrix} m^2 & n^2 & 2mn \\ n^2 & m^2 & -2mn \\ -mn & mn & m^2 - n^2 \end{bmatrix}
$$

where $m = \cos\theta$

$n = \sin\theta$

The lamina stress-strain relationship with respect to the (x, y) axes is given by:

$$\begin{bmatrix} \sigma_{xx} \\ \sigma_{yy} \\ \sigma_{xy} \end{bmatrix} = [T]^{-1} [Q] [T] \begin{bmatrix} \epsilon_{xx} \\ \epsilon_{yy} \\ \tfrac{1}{2}\epsilon_{xy} \end{bmatrix} \tag{2.26}$$

Equation (2.26) essentially describes the behaviour of a unidirectional or bidirectional composite layer, the axes (x, y) of which have an orientation of θ to the principal axes. The theory of laminated structures has been treated elsewhere [2.8].

The modulus of elasticity in the reference axes is given by:

$$\frac{1}{E_{xx}} = \frac{\cos^4\theta}{E_{11}} + \frac{\sin^4\theta}{E_{22}} + \left(\frac{1}{G_{12}} - \frac{2\nu_{12}}{E_{11}}\right) \sin^2\theta \cos^2\theta \tag{2.27}$$

Similarly, Poisson's ratio is given by:

$$\frac{\nu_{xy}}{E_{yy}} = \left(\frac{1}{E_{11}} + \frac{1}{E_{22}} - \frac{1}{G_{12}}\right) \sin^2\theta \cos^2\theta - \frac{\nu_{12}}{E_{11}}(\cos^4\theta + \sin^4\theta) \tag{2.28}$$

Also, the shear modulus G_{xy} is given by:

$$\frac{1}{G_{xy}} = 4\left(\frac{1}{E_{11}} + \frac{1}{E_{22}} + \frac{2\nu_{12}}{E_{11}}\right) \sin^2\theta \cos^2\theta + \frac{1}{G_{12}}(\cos^2\theta - \sin^2\theta)^2 \tag{2.29}$$

These equations have been derived by Calcote [2.8]

2.6.2 Homogeneous isotropic material

The stress-strain relationship for a composite consisting of randomly orientated fibres in a matrix, may be established by assuming the laminae to be homogeneous isotropic materials in a state of plane stress. The assumption is similar to that made for homogeneous orthotropic materials.

The Hooke's law relationships are:

$$\sigma_{11} = \frac{E}{1 - \nu^2}(\epsilon_{11} + \nu\,\epsilon_{22}) \tag{2.30}$$

$$\sigma_{22} = \frac{E}{1 - \nu^2}(\epsilon_{22} + \nu\,\epsilon_{11}) \tag{2.31}$$

$$\sigma_{12} = \frac{E}{2(1+\nu)} \epsilon_{12} \qquad (2.32)$$

Or in matrix form:

$$\begin{bmatrix} \sigma_{11} \\ \sigma_{22} \\ \sigma_{12} \end{bmatrix} = \begin{bmatrix} Q_{11} & Q_{12} & 0 \\ Q_{12} & Q_{22} & 0 \\ 0 & 0 & Q_{66} \end{bmatrix} \begin{bmatrix} \epsilon_{11} \\ \epsilon_{22} \\ \epsilon_{12} \end{bmatrix} \qquad (2.33)$$

where $\quad Q_{11} = Q_{22} = \dfrac{E}{1 - \nu^2}$

$$Q_{12} = \frac{\nu E}{1 - \nu^2}$$

$$Q_{66} = \frac{E}{2(1+\nu)} = G$$

The above implies that only two elastic constants are required to characterize an isotropic material under plane stress.

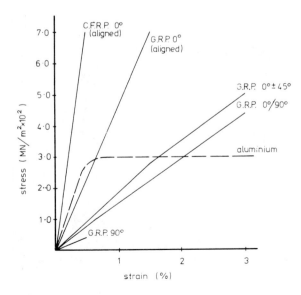

Figure 2.4 Typical stress-strain relationships for various arrangements of fibres in a matrix

2.7 Design criteria for fibre reinforced matrix materials

Safe working stress levels in fibre reinforced matrix materials must be consistent with the known characteristics of the individual components. Figure 2.4 shows typical tensile stress-strain relationships for various arrangements of fibres in a matrix and illustrates the limited strain capability of fibre reinforced matrix material compared with that of aluminium.

In design, the ultimate strength of a fibre/matrix composite will be directly related to the individual components; consequently, the ultimate stress of the fibre must be chosen on the basis of the statistical parameters of strength distribution. Figure 2.5 shows a typical strength distribution for E glass fibres and emphasizes the above requirement. To improve the shear strength

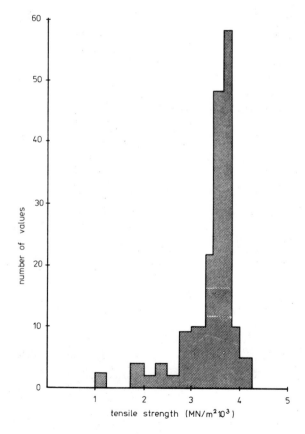

Figure 2.5 Typical strength distribution for E glass fibres

between individual laminae of the composite, a high fracture strain matrix should be used.

It is worth mentioning here that deflection is the limiting criterion for the design of an all glass reinforced plastics structure for the construction industry, and consequently the composite material stress level is generally low.

2.8 Macro analysis of a fibre matrix composite stress distribution

The potential advantages of fibre reinforcement were established on the basis of simple reinforcement configurations subjected to uniaxial loads. The interpretation of these results in the more complex stress situations which pertain to real engineering problems is no simple matter and this difficulty is reflected in current design procedures which with few exceptions refer to

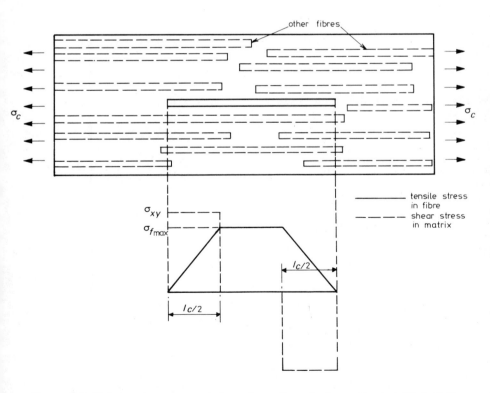

Figure 2.6 Schematic representation of an aligned, discontinuous fibre composite subjected to an axial stress; stress distribution at failure is shown

H.L.Cox.   1952

Cox considered an elastic fibre completely
bonded into an extensive elastic matrix which
is subjected to a uniform stress along the
direction of the fibre axis. Results have been
derived for 2-& 3-dimensional configurations

J.O.Outwater. 1956

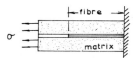

Outwater assumed that the maximum value of
the interface shear stresses which control the
transfer of load to the fibre is a function of
the interface pressure developed by differential
shrinkage of the matrix onto the fibre.

N. F. Dow.   1963

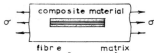

Dow. Axial load is applied to a cylindrical
matrix containing fully bonded elastic fibre. The
solution implies that initially straight lines remain
straight after deformation. 2-& 3-dimensional
configurations may be analysed.

W. B.Rosen. 1964

Rosen assumed perfect bonding at the interface,
the fibre and composite material being subjected
to tensile stress only. The matrix only
carries shear stress

A. Kelly. and W.R.Tyson.
1965

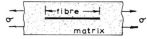

Kelly & Tyson used a model consisting of an
elastic fibre in a plastic matrix which yields
according to the Tresca criterion

Figure 2.7 Isolated-fibre models considered by various authors in developing
approximate analysis

homogeneous biaxial or uniaxial stress fields. The wide range of possible material parameters makes the provision of a comprehensive stress analysis and hence design data extremely difficult, and inevitably attention has been focused upon the analysis of a limited range of fibre arrays. These arrays have been illustrated in Figure 2.6.

When there is no continuous loading path the matrix plays an important part in the reinforcement by transferring load from one fibre to another. Several theories (Figure 2.7) have been presented on the longitudinal stress distribution along the fibre matrix interface, and in the matrix field for

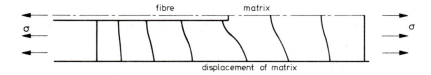

Figure 2.8 Displacement space figure (*After Carrara and McGarry* [2.9])

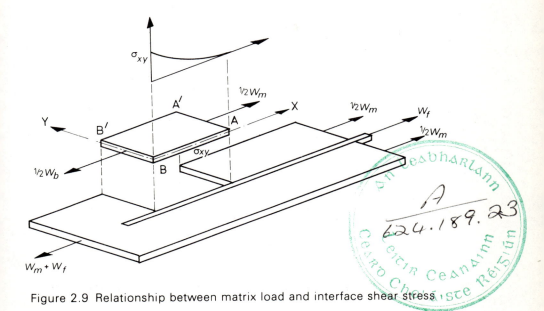

Figure 2.9 Relationship between matrix load and interface shear stress

discontinuous fibres. From these theories the authors have predicted failure mechanisms for fibre reinforced materials.

If the fibres have an elastic modulus greater than that of the matrix, they inhibit its free deformation. This is shown schematically for an isolated fibre in Figure 2.8 after Carrara and McGarry [2.9] and it results in a strain perturbation in the matrix. The load is transferred from the matrix by shear stresses along the edges of the fibre and by direct stress at the fibre ends.

The axial stress in the fibre rises to a maximum value equal to that for a continuous fibre σ_f while the shear stresses, high at the fibre ends diminish to zero; this is illustrated in Figure 2.9. The transfer or critical length is the length of fibre necessary for the continuous fibre stress σ_f to be attained.

2.8.1 Unidirectional fibre array

In the following discussions only unidirectional discontinuous fibre arrays will be considered. One of the most sophisticated analyses of the elastic

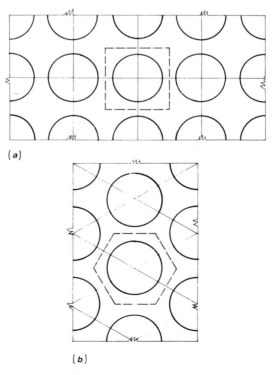

(a)

(b)

Figure 2.10 Array of unidirectional cylindrical fibres; (a) rectangular array; (b) hexagonal array

moduli of composites reinforced by a regular array of unidirectional cylindrical fibres is that due to Pickett[2.10]; the arrays are shown in Figure 2.10. In the case of the rectangular fibre array, he showed that a maximum of 9 elastic constants is required to express the relationship between the 6 macroscopic elastic stress components and the corresponding strains.

$$
\begin{bmatrix} \sigma_{xx} \\ \sigma_{yy} \\ \sigma_{zz} \\ \sigma_{yz} \\ \sigma_{xz} \\ \sigma_{xy} \end{bmatrix} = \begin{bmatrix} g_{11} & g_{12} & g_{13} & 0 & 0 & 0 \\ g_{12} & g_{22} & g_{23} & 0 & 0 & 0 \\ g_{13} & g_{23} & g_{33} & 0 & 0 & 0 \\ 0 & 0 & 0 & g_{44} & 0 & 0 \\ 0 & 0 & 0 & 0 & g_{55} & 0 \\ 0 & 0 & 0 & 0 & 0 & g_{66} \end{bmatrix} \begin{bmatrix} \epsilon_{xx} \\ \epsilon_{yy} \\ \epsilon_{zz} \\ \epsilon_{yz} \\ \epsilon_{xz} \\ \epsilon_{xy} \end{bmatrix}
$$

For a hexagonal or square array the number of independent constants is reduced to 6 with:

$$g_{11} = g_{22}, \quad g_{13} = g_{23}, \quad g_{44} = g_{55}$$

It is assumed that the constituents of the composite are elastic and isotropic; and that the material is subjected to a homogeneous state of stress. Solutions of the equations of elasticity are written in cylindrical co-ordinates and it remains to satisfy the continuity conditions at the fibre-matrix interface and at the boundaries of the typical cell. This may be achieved by 'point matching' in which continuity is satisfied at arbitrarily selected points on the boundary or 'least squares' in which the squares of the errors integrated along the boundary is minimized by variation of the coefficients of a Fourier series.

Much of the recent interest in the development of methods for predicting deformation behaviour under more complex conditions may be attributed to Cox[2.11] who derived elastic constants for planar mats of fibres subjected to both normal and shear stresses applied in the plane of the mat.

A review of available methods for analyzing laminated composites has been prepared by Jones [2.12]. He has drawn attention to the effect of voids and uneven filament spacing and concludes that the simplest models of displacement behaviour yield results which are adequate for design purposes.

2.8.2 Analysis of stresses within an undirectional fibre-matrix composite

The analysis of representing fibre composites in which one or more of the fibres have been broken, either during fabrication or because of prior loading conditions is generally made on two-dimensional models which attempt to

C

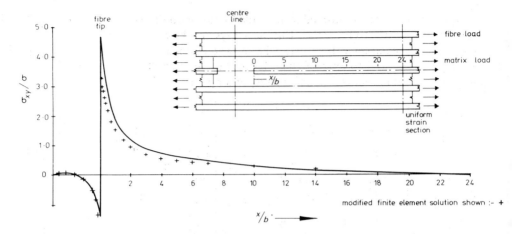

efigure 2.11 Interface shear stresses along centre fibre—fibre end spacing 6 fibre widths.

simulate a broken fibre surrounded by continuous fibres. Fibre breaks during fabrication may result in the gap between the fibre ends being filled with matrix material. A similar geometry can arise from discontinuous fibres being aligned longitudinally and then butted together during fabrication of the composite. Figure 2.6 represents an aligned discontinuous fibre composite.

Because of the difficulties of analysis, the analytical or experimental models are essentially very simplified, being two-dimensional and of regular geometry. Perfect bond between fibre and matrix is assumed. It is also assumed that a continuum mechanics approach is valid, this implies that the fibre and interfibre dimensions are not so small that dislocation and grain boundary movement have a significant effect on the stress field. Both fibre and matrix are taken as homogeneous and isotropic.

Two methods of analysis, one analytical and one experimental, which have been employed to determine the stress distribution in a loaded, idealized fibre matrix composite with various fibre arrangements and with certain restrictions, appear to offer great potential in terms of detailed information. One involves the application of finite element technique and the other the analysis of photoelastic models.

A typical distribution of the interface shear stress for the fibre width gap spacing normalized with respect to σ, the nominal matrix tension, plotted against the distance from the fibre tip is shown in Figure 2.11.

Figure 2.12 (Opposite) Load transfer to fibre A for: (a) a fibre gap spacing of 6 fibre widths; (b) a fibre gap spacing of 1 fibre width

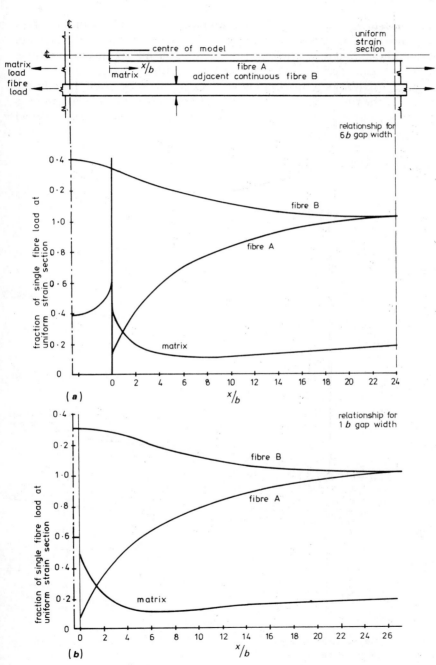

(a)

(b)

A typical load transfer from the discontinuous fibre to the continuous fibres along the critical length is shown in Figure 2.12. These plots show the load in the continuous and discontinuous fibres and in the matrix enclosed by these fibres for a gap spacing of one and six fibre widths.

Figure 2.13 shows a distribution of the axial stresses in the fibre in the vicinity of the fibre tip for a fibre volume fraction of 0.33 and end gap of $\frac{1}{8}$ fibre width. There is a high stress in the fibre right to the tip for this small end

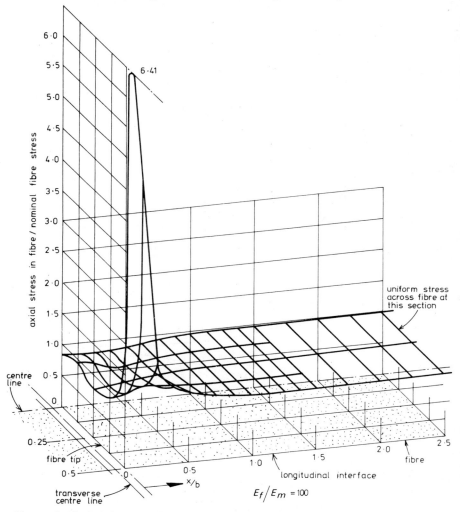

Figure 2.13 Axial stress distribution in fibre in vicinity of fibre tip: fibre volume fraction 0.33; end gap spacing 0.125 fibre width

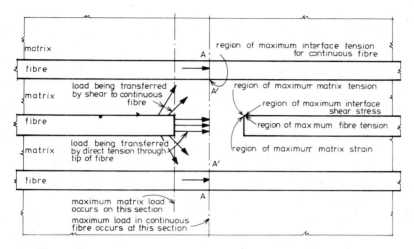

Figure 2.14 Location of critical stresses in multi fibre model

gap but as the fibre tip spacing increases the axial stress decreases.

Figure 2.14 shows the location of the critical stresses in a unidirectional multi-fibre model and also indicates the way in which load is transferred from the broken fibre. It will be seen from the figure that the maximum load in the continuous fibre will be at the centre line between tips of the broken fibres (Section A-A). The maximum interface shear stress will occur in the vicinity of the tip and the maximum interface tension will be located at the corner tip of the fibre.

The importance of the interface geometry in determining the magnitude of the critical elastic stresses in the matrix is illustrated in more detail in references [2.13] and [2.14]. It is also shown in these references that the overall rate of load transfer to the fibre is relatively insensitive to the magnitude of the maximum stresses induced in the fibre tip. The critical length of the discontinuous fibre in the multi-fibre array is found to be dependent upon the fibre volume fraction, the fibre and matrix elastic modulus ratio, and the end gap spacing of magnitude less than one fibre width.

References

2.1. SHAFFER, M., 'Stress-strain relationships of reinforced plastics parallel and normal to their internal filaments', *J.AIAA*, **2**, 1965, pp. 348-352.

2.2. HASHIN, Z. & ROSEN, B.W., 'The elastic moduli of fibre reinforced materials', *J. Appl. Mech.*, **31**, 1964, pp. 223-232.

2.3. SENDECKYJ, G.P., 'Longitudinal shear deformation of composites', *J. Comp. Matls.*, **4**, Oct., 1970, pp. 500-511.

2.4. DOW, N.F. & GRUNTFEST, I.J., 'Determination of most needed potentially possible improvement in materials for ballistic and space vehicles', General Electric Report, TIS 60, SD 389, 1960.

2.5. ROSEN, B.W., 'Mechanics of composite strengthening', Proc. of ASM Seminar 1964, 'Fibre Composite Materials', 17-18 Oct. 1964, pp. 37-75.

2.6. FOYE, R.L., 'Compressive strengths of unidirectional composites', A.I.A.A. paper 66-143, Third A.I.A.A. Aerospace Meeting, New York, 24-26 Jan. 1966.

2.7. COOPER, G.A. & KELLY, A., 'Role of the interface in fracture of fiber composite materials', ASTM STP 452, 1969.

2.8. CALCOTE, L.R., The Analysis of Laminated Composite Structures, Van Nostrand-Reinhold, 1969.

2.9. CARRARA, A.S. & McGARRY, F.J., J. Comp. Matls. 2, 2, 1968, p. 222.

2.10. PICKETT, A., 'Elastic moduli of fibre reinforced plastic composites', Chapter 2, Fundamental Aspects of Fibre Reinforced Plastic Composites, (ed.) SCHWARTZ, R.T. & SCHWARTZ, H.S., Interscience, 1968.

2.11. COX, H.L., 'The elasticity and strength of paper and other fibrous materials', Brit. Appl. Phys., 3, 1952, p. 72.

2.12. JONES, B.H. 'Predicting the stiffness and strength of filamentary composites for design applications', Plastics and Polymers, 36, 122, 1968, pp. 119-127.

2.13 HOLLAWAY, L., 'Fibre reinforced composites', J. Soc. of Engrs., LXI, 1, Oct.-Dec., 1970.

2.14. MacLAUGHAN, T.F., 'A photoelastic analysis of fibre discontinuities in composite materials', J. Comp. Matls. 2, 1, 1968, pp. 44-55.

3 Structural reinforced plastics, mechanical properties of components and fabrication techniques

3.1 Introduction

Unreinforced plastics exhibit poor mechanical behaviour when under load. Table 3.1 compares the most important mechanical properties of some unreinforced plastics with traditional structural materials. It is seen that the magnitude of most properties of plastics are of similar value to those of timber apart from the modulus of elasticity which is about one third that of timber. Moreover the considerable creep deformation of plastics when under load and the non-linear stress strain behaviour of the material do not rule out its

Table 3.1 Comparison of some of the most important mechanical properties of unreinforced plastics with traditional structural materials

Material properties	Specific gravity	Ultimate tensile strength (MN/m²)	Modulus of elasticity in tension (GN/m²)	Coefficient of linear expansion (10⁻⁶/°C)
Thermosetting				
Polyester	1.28	45–90	2.5–4.0	100–110
Epoxy	1.30	90–110	3.0–7.0	45–65
Phenolic (with filler)	1.35–1.75	45–59	5.5–8.3	30–45
Thermoplastic				
PVC	1.37	58.0	2.4–2.8	50
ABS	1.05	17–62	0.69–2.82	60–130
Nylon	1.13–1.15	48–83	1.03–2.76	80–150
Mild steel	7.8	370–700	210	12–13
Aluminium	2.8	450	70	23
Timber (Douglas Fir)	0.5	74	10	4

utilization as the matrix or secondary phase. To be able to increase the potential of plastics materials high strength, high modulus fibres are used to reinforce the secondary phase. These may be either inorganic or organic fibres.

Effective reinforcement may be obtained by the addition of high strength inorganic fibres. Table 3.2 gives the mechanical properties of the better known fibres from which it may be seen that the ultimate strength and modulus of elasticity of these fibres which would be utilized as the dispersed phase are high compared with those of the matrix.

Table 3.2   Typical mechanical properties of the better known inorganic fibres

Inorganic fibres	Ultimate tensile strength (MN/m²)	Modulus of elasticity (GN/m²)
E glass	3500	73.0
S glass	4900	87.0
Z glass	1650	70.0
Stainless steel	2800	203.0
Asbestos	2100–3500	170.0
Boron	2100	420.0
Carbon fibres	2800	350.0
Sapphire whiskers	28000	2100.0

Table 3.3 gives the mechanical properties of organic fibres. The strength and stiffness of the synthetic fibres are increased considerably by the parallel orientation of their long chain molecules, which is achieved by extrusion and after-stretching during the fabrication process. Synthetic fibres are often used in the form of fabrics and unwoven mats.

Table 3.3   Typical mechanical properties of organic fibres

Organic fibres	Specific weight	Ultimate tensile strength (MN/m²)	Modulus of elasticity (GN/m²)	Elongation at rupture (%)
Natural				
Cotton	1.5	500–880	0.05	7–14
Jute	1.5	460	–	4
Sisal	1.45	850	–	2.5
Synthetic				
Viscose rayon	1.52	290	0.04	20–30
Polyamide-6	1.14	560–820	0.004	10–40
Polyacrylonitrile	1.17	410–560	–	15–20
Linear polyester	1.38	750–1000	0.015	20–25
Kevlar 49	1.45	2700	130.0	2.1

The most widely used reinforcement for plastics used in construction is the E glass fibre (see section 3.3); to a much lesser extent, carbon fibre is used as unidirectional reinforcement placed in the high stress region of the composite. Glass fibre combines good mechanical behaviour with relatively low cost and carbon fibre provides an extremely stiff laminate.

Synthetic and natural fibres such as nylon, rayon, polyester, acetate, Kevlar, asbestos, sisal and jute have been combined with resins to form fibre/matrix composites. These laminates have been used experimentally for special purposes in civil engineering, but none of them has had the same impact on the construction industry as composites manufactured from glass fibre/matrix materials. However, this chapter will be limited to a discussion of E glass and carbon reinforcing fibres in conjunction with the resins described below.

3.2 Resins used in structural plastics

Plastics consist essentially of two elements: carbon and hydrogen. The existence of the vast number of carbon compounds which are found in nature or which can be synthesized can be accounted for by the fact that carbon possesses four valencies and has the ability to satisfy any of these valencies by combining with other carbon groups or other groups of atoms having an unsatisfied valency. Plastics are organic materials with very high molecular weights constructed from simpler repeating units under suitable conditions of temperature and catalytic action. The 'building up' of plastics from simpler units is called polymerization and plastics are known as high polymers (see definition in Appendix, p. 217).

Because of their amorphous nature, most plastics are characterized by low elastic moduli (see section 3.1). They all have low specific gravities and high coefficients of thermal expansion (see section 5.3).

Table 3.4 Grouping for the most common plastics used in the construction industry

Thermosetting	Thermoplastic
Polyester	PVC
Epoxy	Acrylic
Polyurethane	Polystyrene
Phenolic	

Although the number of individual plastics runs into many hundreds, they can be broadly classified into two groups: thermoplastic plastics and thermosetting plastics. Table 3.4 gives the grouping for the most common plastics used in the construction industry.

3.2.1 Thermoplastic plastics

Thermoplastics consist of linear polymer molecules which are not interconnected. The chemical valency bond along the chain is extremely strong, but the forces of attraction between adjacent chains are weak. Because of their unconnected chain structure, thermoplastics may be repeatedly softened and hardened by heating and cooling respectively; with each repeated cycle, however, the material tends to become more brittle. Techniques used for moulding thermoplastics include injection and compression moulding, extrusion, cold-drawing and casting from solution. Because of their nature these materials are not generally used for load-bearing purposes.

3.2.2 Thermosetting plastics

Thermosetting plastics are in a liquid or plastic stage once, and then harden irreversibly; this chemical reaction is known as polyaddition, polymerization or curing, and on completion of this reaction the material cannot be softened or made plastic again without change of molecular structure. Unreinforced thermosetting plastics are not suitable for structural elements but when reinforced with fibres they form a composite which has adequate load-bearing properties. Because of their cross-linked molecular structure they are more resistant to heat, chemicals and solvents than thermoplastics.

By far the most important plastics used in the construction industry are the polyester resins. Unsaturated polyester resins are manufactured by reacting together a dihydric alcohol and a dibasic acid, either or both of which contain a double bonded pair of carbon atoms; they are then reacted with styrene at the fabrication plant. Typical mechanical properties of cured polyester resin without filler are given in Table 3.5; specifications for polyester resin are given in BS 3532: Unsaturated polyester resin systems for low pressure fibre reinforced plastics.

Table 3.5 Typical mechanical properties of cured polyester resin without filler

Specific gravity	1.28
Tensile strength	45–90 MN/m^2
Compressive strength	100–250 MN/m^2
Impact strength	1.8–2.4 kJ/m^2
Modulus of elasticity (tension)	2.5–4.0 GN/m^2
Elongation at break	2%
Hardness (Rockwell M scale)	100–115
Thermal conductivity	0.2 W/m deg C
Coefficient of linear expansion	100–110 $\times$ 10^{-6}/deg C
Shrinkage	0.005–0.008
Water absorption 24 h at 20° C	0.15–0.20%

3.3 Glass fibre reinforcement in structural plastics

Glass fibres are manufactured for the reinforced plastics industry by the rapid drawing of molten glass from an electrically heated furnace continuously and

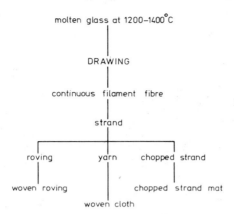

Figure 3.1 Schematic diagram for the manufacture of glass fibre

at high speed through platinum bushings. Its manufacture is shown schematically in Figure 3.1. The filaments cool from the liquid state, at a temperature of about 1200° C, to room temperature in approximately 10^{-5} second. On emerging from the bushings 204 filaments are bundled together and these filaments are bonded to each other by a lubricant or size to reduce the abrasive effect of filaments rubbing against one another.

There are four types of glass fibre:

(a) E glass, of low alkali content is the commonest glass on the market and is used in the construction industry. However, it was not until the advent of polyester resin in 1942 that this glass was used successfully; attempts previously to use E glass in reinforcing thermosetting resins were disappointing [3.1]. Now it is used widely for this purpose, especially with polyester and epoxy resins.

(b) A glass of high alkali content was formerly used in the aircraft industry but is now gradually going out of production. It has been suggested that E glass laminates are better under long term moisture conditions but this is not borne out in practice where it has been demonstrated that A glass reinforced plastics laminates have equally satisfactory service under similar conditions.

(c) S glass has been used by Hawker Siddeley for space research.

(d) Z glass (zirconia glass) has been recently developed in fibre form by Pilkington Brothers Ltd., and the Building Research Establishment. This glass has a high resistance to alkali attack and is used as a reinforcing fibre for cements, mortars and concretes.

Bundles of 204 filaments or multiples of these are called *strands,* and these are usually combined to form thicker parallel bundles called *rovings.* Strands may also be twisted to form several types of *yarns;* rovings or yarns may either be used individually or in the form of woven fabric. A great variety of fabrics is available most of them being plain or square weaves, having identical characteristics in the weft and in the warp direction.

Glass strands for reinforcing thermosetting resins may be used in a number of different forms:

(a) Chopped strands are continuous strands chopped into 50 mm lengths. Due to the method of dispersing these fibres in the matrix, the distribution is generally very uneven and consequently the laminates are not usually manufactured with this form of reinforcement.

(b) Chopped strand mat is manufactured from chopped strands and is probably the most important form of glass fibre reinforcement in present day use. The glass strands are bonded together in a random, two-dimensional manner with a resinous binder and the resulting laminate is assumed, for design purposes, to be isotropic. The strands are normally 50 mm long and there are several different binders which must be compatible with the resins used; binders are chosen according to application. The tensile strength of the laminate will depend upon the amount of glass and the type of resin used, but the chopped strand laminate has exceptionally good interlaminar cohesion and impact strength. Chopped strand mat is considerably cheaper than woven fabric when measured on a weight basis and is invariably used as the reinforcing material in a laminate except when high strength is required, in which case it would probably be replaced by glass cloth.

(c) The manufacturing technique for the production of woven glass cloth is similar to that used in the textile industry. The thinner cloths produce laminates of high tensile strength but lower interlaminar cohesion, and on a weight basis are more expensive to use than the heavier fabrics. The ultimate tensile strength of glass cloth laminates is generally greater than that required in the construction industry.

Woven glass fabrics are expensive for most civil engineering applications and, therefore, more economical forms of glass fibre reinforcement are used for all but the high strength applications.

(d) Woven rovings are used in mouldings and laminates to produce high directional strength characteristics. Unidirectional roving cloths have great strength in one direction; this is achieved by using a high percentage of roving for the warp direction, and a small percentage for the weft direction. Bidirectional roving cloth laminates have high strength properties in two directions at right angles to each other. It is difficult to produce a woven roving laminate with a good surface finish and the interlaminar cohesion between layers is not good. Woven roving can be used in conjunction with chopped strand mat to give bulk and extra strength to laminates.

(e) Surface tissue, which is a thin glass fibre mat bonded with a readily wetted medium, is used when a resin rich surface is required or when the coarse fibre pattern of a chopped strand mat is to be concealed.

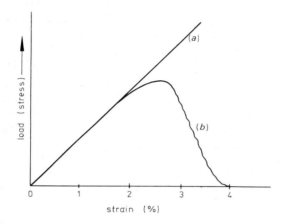

Figure 3.2 Typical o–e curve for glass filament; (a) stress calculated on actual area; (b) stress calculated on total area

Glass fibres under load are elastic until failure and exhibit negligible creep under controlled dry conditions. There is general agreement among various investigators that the modulus of elasticity of a monofilament of E glass is of the order of 73 GN/m². The ultimate fracture strain of the monofilament is about 2.5–3.5%. The stress-strain characteristics of strands have been investigated by Thomas [3.2] and the general form of this relationship is shown in Figure 3.2. The fracture of the strand is an accumulative process of the fracture of its constituent filaments in which the weakest fibres fail first, and the load on the strand is transferred to the remaining stronger fibres, and so on.

3.4 Carbon fibre reinforcement in structural plastics

There are two general sources for commercial carbon fibres: synthetic fibres similar to those used for making textiles, and pitch which is a by-product obtained by the destructive distillation of coal.

The acrylic fibre, PAN (see Appendix, p. 218) is used in all commercial carbon fibre produced in Britain, the development process being patented by the Royal Aircraft Establishment, and several firms are under licence to manufacture these fibres.

The mechanical properties and behaviour of carbon fibre when under load are controlled by its molecular structure; the more highly ordered and defect-free this is, the greater will be the modulus of elasticity and ultimate tensile strength. Certain synthetic fibres have these structural regularities but the formation of carbon fibres requires temperatures in excess of 1000° C at which most synthetic fibres melt and vaporize. However, acrylic fibres and viscose rayon do not melt before decomposition and thus their molecular orientation can be retained during high temperature carbonization.

There are two fundamental types of carbon fibre, the high modulus (type I), and high strength (type II). The Royal Aircraft Establishment, whose initial investigation was into the development of high stiffness fibres, identified these as type I, and the later development of high strength fibres was denoted as type II. The difference between these fibres is in the molecular structure, and to be able to understand how carbon fibres are formed from acrylic precursors, it is important to know their molecular structures and why these lead respectively to high modulus and high strength materials.

It is outside the scope of this book to discuss the chemical structure and manufacturing techniques of carbon fibres; these have been adequately covered elsewhere [3.3], [3.4], and it is sufficient to say that when the macromolecules of the acrylic precursors are examined, two aspects must be considered:

(a) the arrangement of the atoms in the basic unit which is repeated to form the molecules;
(b) the arrangement of these basic units to form a microstructure.

Both types of carbon fibre have the same arrangement of atoms in this basic unit, but their microstructure is different.

The difference between the strengths of type I and type II carbon fibres is not fully understood, but the reason for the variation in modulus is comparatively well known in terms of crystallite orientation. The crystallites in type I fibre are more nearly aligned with the fibre axis than are those in type II fibre; this results in a stiffer structure.

Although carbon fibres have high specific strength and modulus they are not, at present, used very widely in the construction industry in fibre form; this may be due partly to the bonding difficulty and partly to economics. When polyester and epoxy resins are used as bonding agents for the carbon fibre, certain problems that immediately arise are associated with the nature of the bond between the polymer and the fibre and the differences in physical and mechanical properties between them.

Carbon fibres are produced in an inert atmosphere and consequently, because of the nature of carbon, their surfaces are particularly unreactive chemically; this results in poor bonding with the polymers, which in turn, leads to a low interlaminar shear strength. During the initial investigations of this problem, it was thought that a solution could not be found but the situation is readily improved by a method similar to that of activating charcoal, in which atoms taking part in the electron delocalization phenomena will form a chemical bond with the fibres' surface. However, a major disadvantage in undertaking this process is that the surface of the fibres becomes more hydrophilic and in some environments water absorption can take place; however, the uptake of water in a carbon fibre reinforced plastics composite is dependent upon the resin and curing agent used; some resins have a low absorption figure, others have a medium one (most polyesters and epoxies) and a few have a relatively high absorption, e.g. anhydride cured epoxy systems.

If absorption takes place, the modulus of elasticity and the tensile strength of the material are not affected but the flexural and interlaminar shear strengths suffer a 20–25% loss. This situation becomes static after an absorption of about 3%, and no further loss in strength results with increase in water uptake.

Carbon fibre is supplied as a tow containing either 5000 or 10000 individual parallel filaments each of which has a diameter of about 8–9 microns; the continuous filament tow is available in lengths of between about 2000 and 10000 metres on spools or in cakes. In addition to the high tensile strength and high modulus fibres, there is also a high strain fibre. Typical mechanical properties for the three different fibre types are given in Table 3.6.

Table 3.6 Typical mechanical properties for the three types of carbon fibre

	High modulus	High strength	High strain fibre
Specific gravity	1.86	1.74	1.6
Tensile strength (ultimate) (GN/m²)	1.5–2.1	2.4–2.8	1.9–2.6
Modulus of elasticity (tension) (GN/m²)	$3.1–3.45 \times 10^2$	$2.2–2.4 \times 10^2$	$1.75–2.05 \times 10^2$
Shear modulus (transverse) (GN/m²)	280	280	280
Strain at failure (%)	0.6–0.7	0.8–0.9	1.0

3.5 Fibre/matrix composites

As has been pointed out in section 3.2, thermosetting plastics must be reinforced with glass fibres if they are to be used structurally. However hybrid composites of glass/carbon fibres and thermosetting plastics may also be used.

A wide range of composites with varying mechanical properties may be obtained by altering:

(a) the relative proportions of thermosetting plastics and glass fibre;
(b) the type of each component employed;

(c) the fibre orientation within the resin matrix;

(d) the method of manufacture of the composite.

The fibre orientation within the matrix will be dependent upon the type of reinforcement used; consequently, it will be clear that the three main types of composites which can be manufactured will depend upon whether unidirectional strand, bidirectional strand or cloth, or chopped strand mat reinforcement is used.

3.6 Methods of manufacture of G.R.P.

There are various techniques for the manufacture of G.R.P. and these may be considered under two main headings. The first one is the open mould system in which, during the moulding operation, the material is in contact with the mould on one surface only and this is the one generally adopted for civil engineering structural applications. The second one is the closed mould technique in which the composite is shaped between the male and the female moulds. This method is generally employed for the manufacture of small components which are not necessarily associated with the building trades.

3.6.1 Open mould systems

These production methods take full advantage of two important characteristics of polyester resin; it does not require heat or pressure for complete polymerization to occur. There are many variations of the open mould technique, but in the following sections only the principal ones will be given.

(a) Hand lay-up methods

In this technique only one mould is used, and this may be either male or female. Most materials are suitable for mould making, but probably the most common one is G.R.P. A suitable master pattern is prepared and from this G.R.P. moulds may readily be made. Figure 3.3 shows the hand lay-up operation. To prevent bonding of the G.R.P. components, a release agent is applied to the mould and then allowed to dry before any lay-up is undertaken.

The durability of a G.R.P. composite in the construction industry is dependent upon the quality of the surface which is exposed to the atmosphere

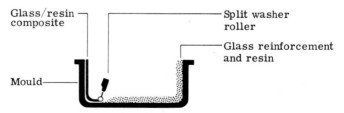

Figure 3.3 Hand lay-up moulding method

and it is necessary to protect the fibres by a resin rich area, known as a gel coat, on the exposed surface of the composite. The function of the gel coat is:

(i) To protect the glass fibre from external influences, the main one being moisture penetration to the interface of the fibre and matrix, with consequent breakdown of the interface bond.

(ii) To provide a smooth finish and reproduce precisely the surface texture of the mould.

The thickness of the gel coat should be controlled to about 0.35 mm and should be uniform throughout. If uneven thicknesses do occur, different cure rates over its surfaces will result, with induced stresses in the resin coat causing crazing. On occasions, a surface tissue mat is used to reinforce the gel coat; this also has the function of balancing the composite throughout its cross-section.

After the gel coat has become tacky but firm, a liberal coat of resin is brushed over it and the first layer of glass reinforcement is placed in position and consolidated with brush and roller. The glass fibre may be in the form of chopped strand mat or woven fabric, which is precut to the correct size. Subsequent layers of resin and reinforcement are then applied until the required thickness of the composite is reached.

(b) Spray-up method

The spray-up technique, as shown in Figure 3.4, is less labour intensive than the hand lay-up method. During the spray-up operation, glass fibre roving is fed continuously through a chopping unit, and the resulting chopped strands are projected on to the mould in conjunction with a resin jet. The glass fibre resin matrix is then consolidated with rollers. The technique requires considerable skill on the part of the operator to control the thickness of the composite and to maintain a consistent glass/resin ratio.

Glass roving — Roving chopper unit
Resin spray gun — Glass reinforcement and resin
Mould

Figure 3.4 Spray-up moulding technique

Although the labour content in this method is less than in the hand lay-up technique, it is considerable in both techniques, and the quality of the finished composite is highly dependent upon the skill of the operators. Curing times are also usually comparatively long. However, the two processes are relatively simple, and low tooling cost can result in considerable versatility from the

D

designers' point of view. Very large mouldings have been produced by these methods.

(c) Filament winding technique

G.R.P. pipes and pressure vessels requiring high mechanical strength in the circumferential and longitudinal directions may be manufactured by passing continuous glass fibre rovings through a bath of activated resin and winding it on to a rotating mandrel. Figure 3.5 illustrates the process, and it will be clear

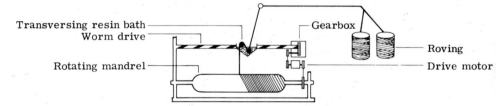

Figure 3.5 Filament winding technique

that the angle of helix is determined by the relative speeds of the traversing bath and the mandrel. The technique can be used in a number of variations of the above and enables a fair degree of automation to be achieved.

3.6.2 Closed mould systems

The production of G.R.P. composites by moulding techniques using matched dies has been undertaken for many years. The hot press moulding technique is usually selected as being the most economical, whilst the cold press technique is an intermediate one between the slower open mould systems which are essentially for large products, and the faster but more expensive hot press moulding system in which long runs of small to medium composites are produced. Resin injection and pultrusion techniques are other systems in the closed mould group.

The matched die processes produce mouldings with a good finish on both surfaces of the composite. Compared with the open mould technique, this system enables a higher glass content to be used in the composite with improved mechanical properties. It also achieves uniform dimensional properties with lower labour costs.

(a) Hot press moulding systems

In the hot press moulding techniques, glass fibre reinforcement and a controlled quantity of hot curing catalyzed resin are confined between heated matched polished metal dies brought together under pressure. The pressures involved vary depending upon the process being undertaken. Figure 3.6 shows a typical hot press mould.

(i) *Pre-form moulding.* In this system chopped rovings are projected on

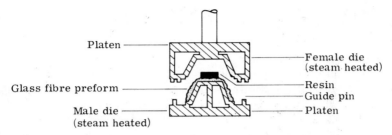

Figure 3.6 Hot press moulding technique

to a rotating fine metal mesh screen which is shaped to the required dimensions. The strands are bound together by spraying the preform with a resinous binder in the form of a powder or an emulsion, and the whole is transferred to an oven at 150° C for 2 to 3 minutes, after which time the preform is ready for the press.

(ii) *Sheet moulding compounds (S.M.C.).* Sheet moulding compound (or prepreg) is a polyester resin based moulding material with E glass fibres, which vary in length and content between 12 to 55 mm and 20 to 35% respectively. They are produced and supplied in the form of a continuous sheet wound into a roll and protected on both sides by sheets of polythene film. The latter are removed before loading into the press. The desired composite shape and rapid cure are obtained by the application of heat and pressure in suitable tools and, as the material flows uniformly to produce a homogeneous composite, complex and deep draw mouldings can be produced. It is necessary to comply with the correct conditions for moulding; these include using a suitable press and a mould designed specifically for the material, charging the mould in the correct manner, and using the optimum temperature, pressure and curing time. Moulding pressures of between 3.5-14 MN/m² and moulding temperatures of between 125°-155° C are generally required; the lower pressures and temperatures are for moulding simple flat shapes and the higher values for the more complex mouldings.

(iii) *Dough moulding compounds (D.M.C.).* Dough moulding compound (or premix) contains an unsaturated polyester resin, an unsaturated cross linking monomer such as styrene, suitable mineral fillers and a fibrous reinforcement which is usually chopped strands. The fibres vary in length and content between 3 to 12 mm and 15 to 20% respectively. Because D.M.C. flows readily, it may be moulded by compression, transfer or injection, and the pressure required to produce a component is relatively low so that large mouldings can be produced without much difficulty. These compounds do not give such high composite mechanical properties as the S.M.C. due to the geometry of the component parts, but they can be used for intricate mouldings. Both S.M.C. and D.M.C. have particular uses in the wide field of appliances and fittings for the building industry, but have not been extensively used in the structural applications of plastics.

(b) Cold press moulding systems

In the cold press moulding technique, matched tools can be used at much lower pressures and temperatures than those for hot press mouldings; consequently, plastics materials can be employed in their manufacture. The pressures may be as low as 100 kN/m². The heat generated during the curing process of the polyester resin is used to warm the tools and to increase the rate of cooling.

(c) Pultrusion process

The pultrusion technique is one in which continuous strands of a reinforcing material are coated with a resin and pulled through a die to form continuous lengths of a desired shape. Thermosetting resins are used almost exclusively with polyester resin comprising the majority of the annual volume of

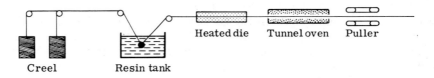

Creel   Resin tank   Heated die   Tunnel oven   Puller

Figure 3.7 Pultrusion technique

pultruded stock. Figure 3.7 shows a diagrammatic representation of the technique. The various stages of the technique are:

(a) resin impregnation;
(b) shape preformer;
(c) cure and tooling;
(d) puller;
(e) cut off.

In the operating process glass fibre rovings are pulled by pullers from creels into a resin bath, where impregnation occurs. The excess resin is removed and the uncured composite is passed through a preformer where the approximate shape is reached. Curing and final shaping is then achieved within a heated die and a flying saw cuts the pultruded composite into the desired lengths.

(d) Resin injection

In this process glass fibre reinforcement is placed in position in the bottom mould and allowed to extend beyond the sides of the moulds. The top mould is then placed in position over the bottom one and secured; thus the reinforcement is held firmly in position whilst the resin is injected. A pressure

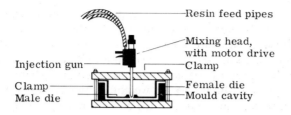

Figure 3.8 Resin injection process

of about 400 kN/m² is sufficient for most mouldings, and when resin seeps through holes specially placed at the highest point of the mould, the resin inlet socket is then sealed. Figure 3.8 illustrates the resin injection process.

3.7 Manufacture of carbon fibre /matrix composites

Probably the future development of carbon fibre reinforced epoxy (or polyester) resin composites (C.F.R.P.) in the construction industry will be in the form of skeletal structures where the members will be manufactured from pultruded or filament wound rods and tubes; an example of this type of structure is given in chapter 7, section 7.4. It is for this reason that only two specific methods for the manufacture of rods and tubes have been discussed.

(a) Pultrusion technique for carbon fibre composites
The technique for the pultrusion of carbon fibre is similar to that for G.R.P. and the diagrammatic representation given in Figure 3.7 may therefore represent it also. It is a sophisticated process for carbon fibre composites explained in more detail in reference [3.5].

(b) Filament winding for carbon fibre composites
The process is similar to that described for glass fibre reinforcement, and consists of winding continuous carbon fibre tows over a mandrel in a machine controlled operation. The process is complicated by the curing operation at elevated temperatures; the technique is also explained in more detail in reference [3.5].

3.8 Categories of fibre /matrix composites

Of the two major components of glass reinforced plastics composites, glass fibres are considerably stiffer and stronger than the resin matrix as has been shown in sections 3.1, 3.2 and 3.3. The mechanical properties of the components are therefore highly dependent upon the content and properties of the reinforcement and upon the arrangement of the fibres in the matrix; the more orderly the arrangement, the more closely the fibres can be packed and, consequently, the greater the proportion of glass that can be used.

It may be deduced from the above that three aspects of a fibre/matrix composite can be broadly considered. Firstly, the condition in which the fibres are randomly orientated in the matrix; in this case, the maximum proportion of the glass in the composite is about 50% by weight. Secondly, the one in which the fibres are orthogonally arranged and where the maximum proportion of glass is about 65% by weight; in this category are the woven rovings and cloth laminates. Finally, the condition in which the fibres are laid in one direction; because they are then packed more closely, the percentage of glass used in the composite can be as high as 90% by weight. However, the practical value is generally nearer 70%.

The mechanical properties for these three categories of glass fibre reinforced plastics material composites will be dealt with in Chapter 4.

References

3.1 WARING, L.A.R., 'Reinforcement', Chapter 10 in *Glass Reinforced Plastics,* (ed.) Parkyn, B., Iliffe, 1970.

3.2. THOMAS, W.F., 'An investigation of the factors affecting the strength of glass fibre strand', *Glass Technology,* **12**, 3, June, 1971.

3.3 GILL, R.M., *Carbon Fibres in Composite Materials,* Iliffe, 1972.

3.4. LANGLEY, M., (ed.) *Carbon Fibres in Engineering,* McGraw Hill, 1973.

3.5. MOLYNEUX, M., 'Workshop practice and processes', Chapter 3 in *Carbon Fibres in Engineering,* (ed.) LANGLEY, M., McGraw Hill, 1973.

4 The mechanical properties and test methods of structural plastics and G.R.P. composites

4.1 Introduction to the mechanical properties

There are two major requirements when designing structural components:

(a) The deformation under load must be within the prescribed functional and aesthetic considerations.
(b) Fracture or rupture should not take place within their scheduled life time.

To satisfy these requirements it is necessary to have information on two particular mechanical properties, namely stiffness and strength. Therefore discussion of the mechanical characteristics will largely revolve around these two properties. Of these two the former is the more important, largely because the stiffness of plastics is lower in relation to strength than for other engineering materials.

4.2 Categories for structural glass reinforced plastics

It was pointed out in Chapter 3 that G.R.P. components may be divided into three categories, and from Figure 4.1 it will be seen that the major difference between these categories is the variation of strength and stiffness with fibre orientation.

Components in which the fibres are randomly orientated will have the same mechanical properties in all directions and are therefore isotropic. Other G.R.P. composites are anisotropic and this property is a consequence of the orthogonal or unidirectional arrangements of the fibres in the matrix; an example of the former is a woven roving laminate and of the latter a unidirectional strand laminate. In the woven roving composite the tensile strength and stiffness is greater in the two orthogonal directions in which the fibres lie than in any other direction, and this results in a particular form of anisotropy known as *orthotropy*. Furthermore, in the unidirectional strand

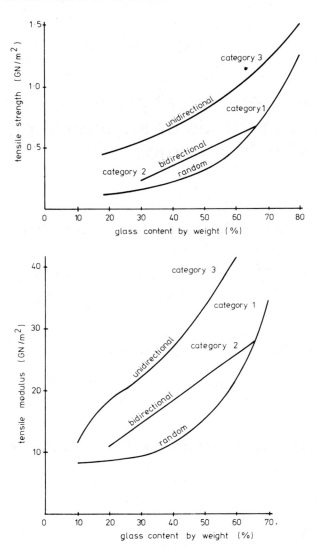

Figure 4.1 Typical tensile strengths and tensile moduli versus glass content

composite, the strength and stiffness is greatest in the direction in which the fibres are aligned so that this also is an orthotropic material. It may be noted in the latter case that in planes perpendicular to the alignment, the strength and stiffness do not vary with direction and, consequently, they are transversely isotropic.

Table 4.1 Typical mechanical properties for glass reinforced plastics composites

Material	Glass content (percent by weight)	Specific gravity	Tensile modulus (GN/m²)	Tensile strength (MN/m²)
Unidirectional rovings (filament winding or pultrusion	50–80	1.6–2.0	20–50	400–1250
Hand lay-up with chopped strand mat	25–45	1.4–1.6	6–11	60–180
Matched dye moulding with preform	25–50	1.4–1.6	6–12	60–200
Hand lay-up with woven rovings	45–62	1.5–1.8	12–24	200–350
D.M.C. polyester (filled)	15–20	1.7–2.0	6–8	40–60
S.M.C.	20–25	1.75–1.95	9–13	60–100

Whenever data are published on the strength and stiffness of G.R.P. composites, there are two important factors which must be understood when interpreting the information given; these are:

(a) the anisotropic nature of the laminate; the data presented generally relate to the maximum values of the composite;

(b) the method of manufacture of the laminate.

Table 4.1 gives the tensile characteristics of the different types of G.R.P. composites and illustrates the above. It is evident that the tensile stress-strain curves of composites will be a function of the component parts; where the glass content is high the curves established in direction of fibre alignment will show a close resemblance to the characteristics of glass and conversely when the glass content is low the matrix characteristics will be reflected.

4.3 Stress-strain relationships for G.R.P.

The uniaxial tensile stress-strain curves for G.R.P. composites are commonly obtained at constant rates of elongation; often following the test procedures prescribed in national standards such as BS 2782 [4.1] or the corresponding standard of the American Society for Testing and Materials [4.2]. (The results in Table 4.1 conform to the above British Standard.) At relatively low strains the stress-strain curves exhibit a linear relationship; these curves are established in a short period of time, usually of the order of a few minutes. At higher levels of strain the slopes of these curves are generally less than their initial values and the change occurs at a point which is often referred to as the knee in the stress-strain curve. The knee is associated with crazing or cracking of the matrix material and once this point has been passed, the composite is permanently damaged and, although it is still capable of undergoing considerable strain deformation, it will not return to its original dimensions when the load causing strain is removed. This characteristic is more noticeable with higher fibre contents and the position of the knee on the stress-strain curve is dependent upon the stiffness of the matrix; for very flexible matrices, cracking may be eliminated altogether.

The apparent linearity of the stress-strain curve before the onset of permanent damage does not imply that the composite is Hookean in nature, but that the method of testing may not be applicable. It will be realized that the plastics component of the composite is a viscoelastic material and that the deformational behaviour of it is highly dependent upon the way in which it is subjected to stress and strain. Glass on the other hand is linearly elastic; consequently, the composite itself will exhibit viscoelastic properties, and under conditions of testing, other than those laid down in BS 2782, the stress-strain behaviour will appear different.

The mechanical behaviour of G.R.P. might be more realistically established by applying constant loads over longer periods of time; these investigations may be defined as creep tests (see section 4.4). These tests produce curves of elongation against time at different stress levels and, although they are not able to produce data which may be converted directly to stress-strain curves, as shown in references [4.3] [4.4] constant time sections through families of such creep curves have been used to produce isochronous stress-strain curves (see section 4.4.3).

It should be realized, however, that the information given in manufacturers' handbooks on the deformational behaviour of plastics is generally derived from the constant rates of elongation tests.

4.4 The deformational characteristics of G.R.P.

There are at least three different ways in which uniaxial tension can be applied to a specimen. These correspond to:

(a) a constant rate of strain;

(b) a constant rate of stress;
(c) a constant stress.

These three methods of testing have been well documented in references [4.5] and [4.6] and only the essential steps will be explained here.

4.4.1 Constant rate of elongation test
The constant rate of elongation test approximates closely to the constant rate of strain test as prescribed by BS 2782 [4.1]; these tests lead to secant and tangent moduli values which vary significantly with differing strain rates. It is, therefore, necessary to define the rate of strain if results are to be intelligible.

4.4.2 Constant rate of loading test
This type of test procedure is relatively uncommon because of the difficulty of applying constant rates of loading (i.e. constant rate of stressing) using standard testing equipment, but if used it is necessary to define the rates of stress used.

4.4.3 Constant load tests (creep tests)
In this case, the tests amount to a series of different constant stress levels applied over varying time intervals and correspond directly to those met in many structural applications of plastics. These tests can cover long as well as short periods of loading and therefore the results are to be preferred to the above two cases when structural design of viscoelastic materials is being undertaken. Greater time and effort are required to obtain the values compared with the test procedures laid down in BS 2782; however, there has been a move towards creep tests to obtain the deformational behaviour of plastics for design purposes. This is exemplified by the recommendations of the British Standards Institution on plastics design data [4.7]. It will be realized that creep tests do not immediately provide isochronous stress-strain curves but these may be obtained by taking constant time sections through creep curves.

The results for idealized linear viscoelastic plastics corresponding to the above three sets of conditions are shown in Figure 4.2. From these it is possible to derive constant time sections through stress-time and strain-time curves to give isochronous stress-strain curves. In the case of a linear viscoelastic plastics material the stress-strain curves will generate straight lines, but for real plastics which are non-linear viscoelastic materials, the curves are as shown in Figure 4.3 which gives a typical pattern of the behaviour.

Currently, the isochronous stress-strain curves are often used to define the behaviour of thermoplastics and, although thermosetting plastics are not at

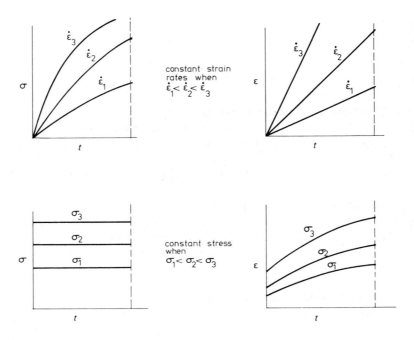

Figure 4.2 Stress-time and strain-time curves for a linear viscoelastic material

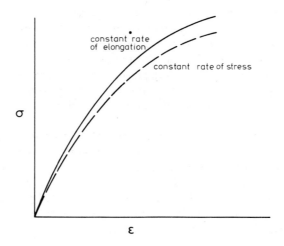

Figure 4.3 Typical isochronous stress-strain curves for a non-linear viscoelastic material

present generally defined by these curves, it may be deduced that their general response will be similar. Furthermore, as the viscoelastic nature of glass reinforced plastics is dependent upon the characteristics of the plastics matrix, it may also be deduced that the stress-strain curves for this composite will produce similar responses to that of the matrix.

The importance of the isochronous stress-strain curves lies in the fact that the slopes obtained from the three methods of test loading are different and that they decrease in the order: constant stress rate, constant strain rate and constant stress.

The constant rate of elongation test makes plastics appear stiffer than they are under the constant load tests and, therefore, the information it will provide on the stiffness of plastics is over optimistic. It may also be observed that the constant load or creep tests relate directly to civil engineering structural problems under long term loading, but the stiffness of components under short term loading will be under estimated by this test procedure and will err on the side of safety.

Special apparatus has been developed at I.C.I.[4.8] to undertake isochronous stress-strain investigations in order to obtain creep characteristics of thermoplastics. Clearly these test procedures should be utilized to obtain stiffness and creep characteristics of G.R.P. components under long term loading but, because of the higher load carrying capacity of G.R.P. composites, modifications to Turner's original apparatus [4.8] would be required. Alternatively, a bending test procedure could be used for this study, thus making it possible to use larger specimens and higher loads with a much simplified apparatus; Mucci and Ogorkiewicz have given details of a suitable arrangement [4.9 and 4.10]. Although many testing laboratories still undertake the more *ad hoc* testing procedures, the unreliability of the creep results and the long term modulus of elasticity given by these clearly indicate that the isochronous stress-strain characteristics of G.R.P. should be obtained. It does, however, require considerably more time and effort to produce these curves compared with those obtained under the test procedures laid down in BS 2782; a further requirement is that the tests should be performed at known temperatures and relative humidities. There is a certain justification for adopting BS 2782 or A.S.T.M. Part 36, when the superimposed loads, applied to civil engineering G.R.P. structures, are of short duration (e.g. wind and snow); the results of the test will underestimate the stress strain relationship which will be a function only of ageing effect of the material.

4.5 Recommendations of BS 4618 for the time-dependence of plastics

At this point it would be helpful to consider briefly the recommendations for

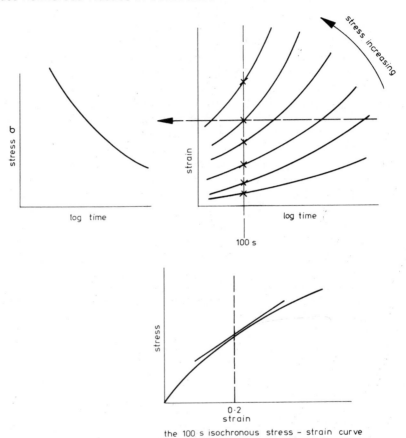

the 100 s isochronous stress – strain curve

Figure 4.4 Creep curves—isochronous stress-strain curve (derived from BS 4618)

creep tests of thermoplastic materials as laid down in BS 4618. In this Standard, constant load tests are defined under controlled conditions for the following durations: 60 s, 100 s, 1 h, 2 h, 100 h, 1000 h, 1 year, 10 years and 80 years. For any material a family of creep curves may be obtained by varying the stress as shown in Figure 4.4. From these curves, isochronous stress-strain curves may be drawn, each corresponding to a specific loading duration. Thus a 100 s isochronous stress-strain curve implies that the total strain at the end of 100 s has been plotted against the corresponding stress level, and it will be seen that the slope of this curve is not constant. Consequently, it is necessary to specify at which point on the curve the slope has been determined and this slope is then defined as the creep modulus. It will have two prefixes to denote the loading duration and the corresponding strain value (for instance, 100 s, 0.2% strain, tensile creep modulus). This creep modulus will then characterize the deformational behaviour of the plastics material.

4.6 Strain recovery

All plastics materials creep under load, whereas the glass or carbon fibre reinforcements have virtually no creep when under load. Consequently, G.R.P. components will follow the general pattern of creep behaviour of the matrix material.

Provided the stress levels in the components are acceptable, the creep of plastics is recoverable; therefore in cases where loading is essentially of an intermittent nature the cumulative effects of creep may safely be ignored.

The rate of strain recovery may be established by additional observations following the creep tests explained above. The amount of experimental work, however, may be reduced by plotting 'fractional recovery' against 'reduced time' where both of these qualities have been defined as [4.11]:

Fractional recovery is the ratio: strain recovery to the creep strain immediately before load removal.

Reduced time is the ratio: recovery time to the duration of the preceding creep period.

4.7 Strength characteristics

As the ultimate tensile strength (defined here as the maximum load which a specimen can support before it breaks, divided by its original cross-sectional area) of plastics materials varies with time, when long duration loading is considered, it is necessary to define this strength in terms of a series of curves of maximum nominal tensile stress versus time. Creep tests to the point of specimen rupture are usually performed to establish these curves and it has been shown [4.12] that in general the tensile strength of G.R.P. decreases with time.

4.8 Time, temperature and humidity effects

As reinforced plastics creep less than unreinforced plastics it may be deduced that the greater stiffness associated with the reinforced plastics does not deteriorate greatly with time. In some cases the rate of creep is significantly lower, which means that the stiffening effect of the fibres increases as the plastics materials soften with the duration of loading. A similar effect is observed when the temperature of the component is increased. Reinforced plastics soften with a rise of temperature but to a much lesser extent than do unreinforced plastics materials. The implication therefore is that the stiffening effect of the reinforcing fibres increases with temperature until this approaches the glass-rubber transition of the plastics matrix.

As the matrix of hydrophilic plastics is moisture softened, the stiffening effect of the reinforcing fibres increases. This again implies that the stiffness of reinforced plastics, when wet, decreases to a lesser extent than in the unreinforced plastics materials.

It should be noted here that additional factors have to be considered when reinforced plastics are in contact with liquids. Unless adequate attention is paid to detail in the manufacture of the composite a breakdown of bond between fibres and matrix could take place, and this would considerably reduce both the strength and stiffness of the composite.

4.9 Impact behaviour

It is desirable to estimate the probability of failure of components when they are subjected to loads which are applied suddenly. However, impact strength is not a well defined physical characteristic of a solid material and it cannot be derived from other basic properties. One property which does give information on impact strength is the ability to absorb energy. Stress-strain diagrams obtained at the constant rate of strain allow estimates of impact strengths to be made for different composites by comparing the areas under the load deflection curves; these areas indicate the work to failure of the specimen. Therefore, impact behaviour can only be assessed by performance tests, which are nevertheless useful for comparing different plastics. British Standard 4618 [4.13] recommends a range of tests covering impact testing of plastics at various temperatures as well as practical tests on finished products. Berry [4.14] has indicated some practical tests on full scale specimens and structures.

As with other mechanical properties of G.R.P., there is little information yet on the impact resistance of this material. Data so far obtained are related generally to thermoplastic materials and indeed impact testing procedures have been developed specifically for these; however, it may be deduced that the impact response of thermosetting plastics will be similar to that of thermoplastic plastics and the test procedures adopted for these are also relevant to G.R.P.

The most convenient method of impact testing is the simple pendulum based on the Charpy, the high speed tensile, and the Izod configurations; this method gives a direct indication of the energy consumed in fracturing the specimen. The Charpy test enables a range of notch geometries, including un-notched, bluntly-notched, and notched specimens, to be investigated, in order to highlight the two different components of the energy; these are crack initiation and crack propagation energies and they are collectively referred to as impact strength. The relative magnitudes of these two components vary in different plastics; the former indicates the relative resistance to damage by a blow, and employs an un-notched or bluntly notched specimen, and the latter signifies the shattering by crack propagation using a notched specimen.

The factors which influence the impact energy are temperature, speed of impact, stress concentration, anisotropy and thickness effects.

4.9.1 Charpy test

The basic principle of this type of test is to allow a pendulum of known energy to strike a rectangular specimen at the lowest point of its swing. Test methods 306D and 306E described in BS 2782 [4.15] are for un-notched and notched

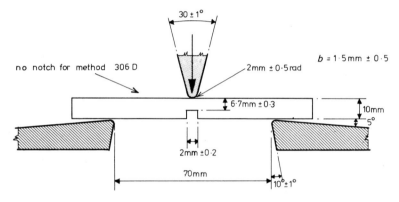

Figure 4.5 Specimen mounting for Charpy test (BS 2782)

specimens respectively and have been formulated from continental methods which have been standardized in I.S.O. Figure 4.5 shows the specimen mounting for BS Charpy test.

The impact resistance of un-notched specimens is calculated from the energy absorbed when the pendulum breaks the mounted specimen; this energy is divided by the cross sectional area of the specimen (method 306D). In the notched method, the energy absorbed at break is divided by the cross-sectional area of the specimen at the root of the notch (method 306E).

4.9.2 High speed tensile impact

This method is also based upon a pendulum machine and has been standardized in A.S.T.M. D1822-68 [4.16]. A diagrammatic representation of a typical high speed tensile impact tester is shown in Figure 4.6. The specimen is fixed at one end of the 'head' of the pendulum, the chuck being so designed as to pass unhindered through the arrester of the machine. The load is applied through the trailing grip which has projecting-pins on each side so that it is arrested by a pair of anvils, one on each side of the grip. In the A.S.T.M. test

E

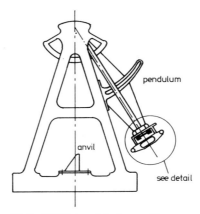

160 Nm AVERY impact testing machine

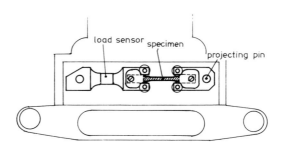

tensile specimen, grips and loadbearing assembly

Figure 4.6 Typical high speed tensile impact tester (*By permission of the Director, The Procurement Executive, Ministry of Defence, E.R.D.E.*)

the velocity of fracture is 3444 mm/s and is achieved by a fall height of 609.6 mm.

A more sophisticated approach adopted by some research laboratories is to measure the load applied to the specimen by means of a load sensor in series with the loading grip, the transient output of which is recorded by an oscilloscope camera system. The high speed tensile impact tester is described in greater detail by Westover and Warner [4.17].

4.9.3 Izod test

The Izod testing machine is essentially a pendulum falling through a known height and striking a standard specimen at its lowest point of swing; the height to which the pendulum continues to swing is then recorded. This method is adequately described in most text books on materials and is fully detailed for plastics materials by Ives, Mead and Riley [4.18]; the test procedure is laid down in BS 2782 : 1970 [4.15].

4.9.4 The variable height impact tester

This tester was originally developed in the U.S.A. [4.19] for 1.5 mm thick sheet material. The impact tester is shown in Figure 4.7. In practice, the failure of a composite is likely to result from a direct blow delivered by the apparatus, but the Building Research Establishment (BRE)[4.20] have found that with sheet material of the order of 3 mm thick the results are less satisfactory than those for material of 1.5 mm thickness and modifications to the apparatus are being investigated.

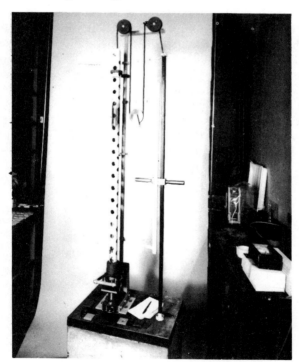

Figure 4.7 Variable height impact tester

4.10 Impact resistance of G.R.P.

Compared with unreinforced plastics, G.R.P. shows improved resistance to impact blows. Values of impact strength as given by the notched Izod test and presented by Scott Bader [4.21] have been reproduced and compared in Table

Table 4.2 Typical impact strength of G.R.P. notched composites (As given by Scott Bader & Co., [4.21])

Impact strength	Chopped strand mat	Woven rovings	Continuous rovings
Izod unnotched (k J/m²)	75.0	125.0	250.0

4.2; the comparisons cover various glass reinforcements. Bader, Bailey and Bell [4.22] have given comparisons of un-notched G.R.P. and C.F.R.P. components (Table 4.3) and have shown that the impact resistance of unidirectional G.R.P. is higher than that of either type I or type II carbon fibre. The reason for this is the extensive delamination which occurs in the G.R.P. composite compared with that in a C.F.R.P. composite; the process of delamination tends to absorb the force of the impact.

The adhesion between the fibre and epoxy matrix in a C.F.R.P. composite is considerably better than that in a G.R.P. one, and at failure due to impact a 'brittle' crack propagates through the composite rather than causing delamination at the fibre matrix interface; consequently, the total energy to destruction in this composite is lower than in a G.R.P. one. As is shown in Table 4.3 a lower impact resistance is obtained on the composites with treated carbon fibre surfaces than on those with untreated surfaces; the function of the treatment is to improve the bond between the fibre and matrix. This is consistent with the above result and explanation.

Table 4.3 Comparison of C.F.R.P. and G.R.P. components of fibre/matrix ratios by weight (w_f) of 0.4 and 0.6

Fibre type	Modulus of elasticity in tension (GN/m²)		Ultimate tensile strength (GN/m²)		Impact resistance unnotched (kJ/m²)	
w_f	0.4	0.6	0.4	0.6	0.4	0.6
I	130	160	0.45	0.60	26	25
I s	130	180	0.55	0.75	10	12.5
II	70	110	0.75	0.15	40	57
II s	75	125	0.95	1.40	35	48
G.R.P.	25	42	0.60	0.85	85	95

Type I and II carbon fibres are the high modulus and the high strength fibres. The suffix s denotes the surface treatment specimens (Morganite Modmor Ltd. proprietary treatment). G.R.P. specimens are manufactured from unidirectional E glass fibre. The resin system was an epoxy based on Shell Epikote 828, DGEBA.
(Results have been abstracted from Bader, Bailey and Bell [4.22])

G.R.P. has been found to be relatively notch sensitive; it follows, therefore, that the performance of a G.R.P. composite in service depends upon the level of cracks, notches and other local defects.

It is necessary to cover a range of temperatures since the relative impact strength of different plastics can change completely with a range of 50° C [4.23]. Consequently, BS 4618 [4.7] on plastics design data advises covering a range of temperatures of practical interest as well as a number of different notch geometries. References [4.23] and [4.7] above deal specifically with thermoplastics, but again it may be deduced that the general response of thermosetting plastics will be similar. Because of the possibility of anisotropy in strength as well as stiffness, it is advisable to cut specimens in more than one direction. They should also be cut perpendicular to the direction of any molecular or fibre orientation, which will show the material at its weakest.

Probably a more reliable result for impact resistance of G.R.P. is obtained by testing a full scale structure as opposed to small test specimens. The Agrement Board [4.24] specify the following tests for external cladding:

(a) soft body impact test in which a 50 kg sand bag is dropped from a height of 2 metres;

(b) hard body impact test in which a 1 kg steel ball is dropped from a height of 1 metre.

The extent of any damage received during these tests is noted.

It will be clear that the requirements for resistance to impact forces depend upon the position of the panel in the building and the use to which the building is put. For some panels, the requirements are usually higher than those specified by the Agrement Board. It may therefore be necessary to carry out simulative or representative impact tests on a prototype panel.

4.11 Torsion test

This test is probably one of the most difficult to perform because in applying a shear force to a specimen other forms of loading are also unavoidably present. However, torsion testing machines have been developed at the I.C.I. Plastics Division for determining the shear modulus of plastics (mainly thermoplastics). Details of the first machine used are given in reference [4.25]; this was subsequently improved by incorporating the use of air bearings to reduce the machine friction to negligible proportions [4.26]. Both of these machines are limited to small shear strains of approximately 0.2%; consequently the shear modulus can be calculated from the slope of the straight line plot of the angle of twist of the specimen against the applied torque using the standard linear elastic solution to the torsion of prismatic bars.

Specimens may be strained by using an interrupted step-loading technique, analogous to that used in the tensile tests, and by observing the twist at, say, 100 seconds after the application of the load, the 100 seconds isochronous torque versus angle of twist curves are obtained. It should be noted, however, that because of the limitation of the apparatus, this approach may only be used with small cross-section plastics specimens and that the apparatus for the investigation of G.R.P. could be similar but would have to be more robust.

Greszczuk [4.27] has suggested several test methods that can be used to determine experimentally the shear moduli of isotropic and fibre/matrix composite materials. The difference between these methods lies in the experimental complexity, the types of specimen utilized and the accuracy of the results they yield. The test methods which he has mentioned are:

(a) the plate-twist test;
(b) the cylindrical-torsion test;
(c) the Douglas ring test.

Only the Douglas ring test will be discussed here.

4.11.1 Douglas ring test

The Douglas ring test consists of a split ring which is subjected to two equal forces applied in opposite directions and normal to the plane of the ring (see Figure 4.8a). The resultant vertical deflection of the point of application of the load is:

$$\delta = \frac{\pi P R^3}{E I} + \frac{3 \pi R^3 P}{G J} \tag{4.1}$$

where the first term on the right hand side represents the component due to moment deflection and the second term represents shear deflection. The polar moment of inertia J of the ring section (rectangular cross-section) is expressed as $J = \alpha a b^3$ where α is a numerical factor dependent upon the ratio a/b; Table 4.4 gives values of α for various values of a/b. For a ring with a circular cross section, $J = \pi r^4/2$, where r is the radius of the cross section.

Table 4.4   Values of parameter α for various values of a/b

a/b	1.0	1.2	1.5	1.75	2.0	2.5	3.0	4.0	5.0	6.0	8.0	10.0	∞
α	0.141	0.166	0.196	0.214	0.229	0.249	0.263	0.281	0.291	0.299	0.307	0.313	0.333

As the ratio E/G increases, the component of deflection due to bending decreases and that due to torsion increases. For materials in which the E/G ratio is 10, the percentage of deflection due to torsion is 94.7% of the total [4.28] and therefore this test is ideal for the determination of the shear moduli.

It may be seen from equation (4.1), that as well as computing the ring geometry, it is necessary to determine the modulus of elasticity of the material. This may conveniently be found by applying two diametrically opposite tensile forces P that act in the plane of the ring and which are applied at $\pm 90°$

from the ring split (Figure 4.8b). The modulus of elasticity is obtained from
the following equation:

$$E = \frac{3 \pi P R^3}{I \delta} \tag{4.2}$$

where δ = change in ring diameter across the points of application of the
load
R = radius of ring
I = second moment of area of the cross section of the ring.

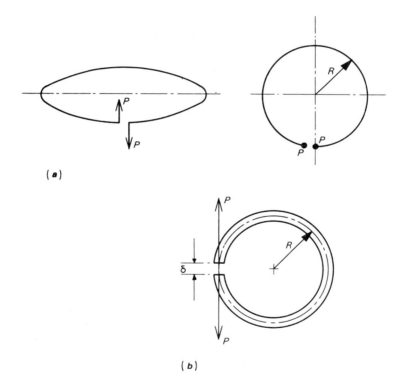

(a)

(b)

Figure 4.8 Douglas ring test procedures; (a) Douglas ring test; (b) procedure
for determining modulus of elasticity

4.12 The viscoelastic behaviour of plastics

It will be helpful to consider the viscoelastic behaviour of materials by means
of models. Models in the form of combinations of springs and dashpots have
been used for many years to describe the mechanical behaviour of materials. It

should be realized, however, that the equation of the theory of viscoelastic behaviour can be derived without reference to the spring and dashpot models, and that the basic function of producing these is to provide a pictorial symbol of the approximate equation which describes the material behaviour, and to show the importance of choosing the relevant test procedures for viscoelastic materials.

4.12.1 Uniaxial stress behaviour of plastics

The total creep curve for plastics under a given uniaxial stress σ and constant temperature T can be divided into five parts; a graph for a highly idealized material is shown in Figure 4.9. It will be realized that some materials do not exhibit a secondary type creep and others do not have a tertiary region; also it

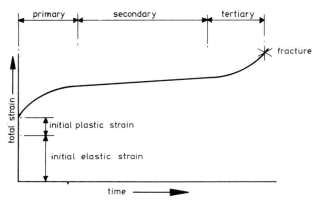

Figure 4.9 Total creep for a highly idealized viscoelastic material (uniaxial stress and temperature constant)

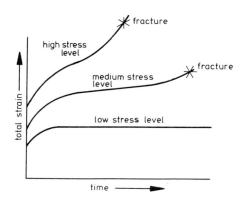

Figure 4.10 Typical family of creep curves (uniaxial stress and temperature constant)

is common to find that in some plastics the tertiary creep is evident only at high stresses. A typical family of creep curves is given in Figure 4.10.

4.12.2 Mechanical models for viscoelastic behaviour

The viscoelastic theory usually considers the combination of two basic types of behaviour; these are Hookean elasticity and Newtonian viscosity. The Hookean elasticity can be represented by a linear algebraic equation and a linear elastic spring, as given in equation (4.3):

$$\sigma = E\,\epsilon \tag{4.3}$$

where σ = applied stress
  ϵ = resulting strain
  E = proportionality constant.

The Newtonian viscosity can be represented by a linear differential equation and a linear dashpot, as given in equation (4.4):

$$\sigma = \eta\,\dot{\epsilon} \tag{4.4}$$

where σ = applied stress
  ϵ = resulting strain
  $\dot{\epsilon}$ = strain rate
  η = proportionality constant.

The value of strain at any time t, can be obtained by integrating equation (4.4).

$$\epsilon = \frac{\sigma}{\eta}\, t \tag{4.5}$$

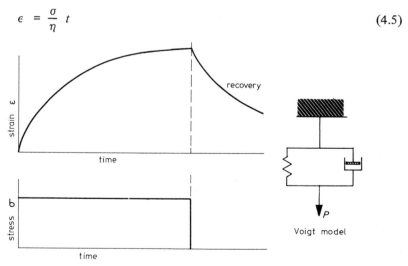

Figure 4.11 Combination of Hookean and Newtonian models to form Voigt model

Creep can be described by combining the Hookean and the Newtonian models in parallel and the total stress is then the sum of the individual components.

$$\sigma = E\epsilon + \eta\dot{\epsilon} \qquad (4.6)$$

This model is known as the Kelvin or Voigt model and is represented in Figure 4.11.

By combining the Hookean and Newtonian models in series (Figure 4.12) the Maxwell model is obtained and the total strain is the sum of the individual components

$$\epsilon = \frac{\sigma}{E} + \frac{\sigma}{\eta} t \qquad (4.7)$$

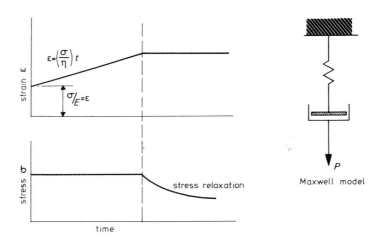

Figure 4.12 Combination of Hookean and Newtonian models to form Maxwell model

A combination of the Kelvin and Maxwell models in series (Figure 4.13) describes initial elastic strain, primary creep, secondary creep and instantaneous elastic recovery, but only specifies a linear relationship between stress and strain and between stress and rate of strain. Although this may be applicable to some groups of materials, it specifies the properties of plastics only very approximately. However, by refining the model systems a more accurate representation may be obtained. It will be seen that the series combination of the Kelvin and Maxwell models describes approximately the idealized constant stress creep curves of Figure 4.9.

Appendix 4.1

The approximately equivalent B.S.S., A.S.T.M. and I.S.O. codes are given below:

B.S.S.	A.S.T.M.	I.S.O
B.S. 2782 'Methods of testing plastics'	A.S.T.M. Standards on Plastics Parts 34, 35, 36	
	Part 34: 'Plastics pipes'	I.S.O. T.C. 138/WG6 Proposed standard A.S.T.M. specification for reinforced plastics mortar, low head pressure pipes
	Part 35: 'Plastics general methods of testing nomenclature paints and applied coatings' Part 36: 'Plastics specifications: methods of testing pipes, film, reinforced and cellular plastics; fibre composites'	At the end of B.S. 2782: 1970 a comparison is made with similar methods standardized or in the course of standardization by T.C. 61 of the I.S.O. In more recently published sections of B.S. 2782, and if applicable, the introduction to each part gives the methods of I.S.O. which are in general agreement with the B.S.S.
B.S. 4618 'Recommendations for the presentation of plastics design data'	No equivalent	No equivalent
B.S. 3532 'Unsaturated polyester resin systems for low pressure fibre reinforced plastics'	No equivalent	No equivalent

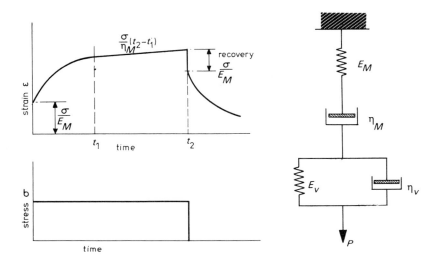

Figure 4.13 Combination of Voigt and Maxwell models in series

References

4.1. 'Methods of Testing Plastics', British Standard 2782, Part 3, British Standards Inst., London, 1970.

4.2. 'A.S.T.M. Standards on Plastics—Parts 34, 35 and 36', American Society for Testing and Materials, Philadelphia.

4.3. OGORKIEWICZ, R.M., (ed.) *Engineering Properties of Thermoplastics,* Wiley, London, 1970.

4.4. OGORKIEWICZ, R.M., *The Engineering Properties of Plastics,* Engineering Guide, 17, Design Council, London, 1977.

4.5. OGORKIEWICZ, R.M., CULVER, L.E. & BOWYER, M.P., 'Stress-strain characteristics of an acetal copolymer under different types of tensile loading', *Plastics and Polymers,* 37, 129, June, 1969.

4.6. OGORKIEWICZ, R.M., CULVER, L.E. & BOWYER, M.P., 'Deformational behaviour of thermoplastics under different types of tensile loading', Proc. 26th Ann. Tech. Conf. S.P.E., New York, May, 1968.

4.7. 'Recommendations for the presentation of plastics design data', British Standard 4618 (Part 1 : Mechanical Properties), British Standards Inst., London, 1970.

4.8. MILLS, W.H. & TURNER, S., 'Tensile creep testing of plastics', *Proc. Inst. Mech. Eng.,* 180, part 3A, 1965-66, pp. 291-302.

4.9. MUCCI, P.E.R. & OGORKIEWICZ, R.M., 'Machine for the testing of plastics under bending load', *Strain Analysis,* 9, 3, 1974, pp. 141-145.

4.10. MUCCI, P.E.R., Diploma Thesis, Imperial College London, 1973.

4.11. RATCLIFFE, W.R. & TURNER, S., 'Engineering design : data required for plastics materials', *Trans. Plastic Inst.,* 34, 111, June, 1966.

4.12. FRINDLEY, W.N., 'The effect of temperature and combined stresses on creep of plastics', 2nd Int. Reinforced Plastics Conf., British Plastics Federation, London, 1960.

4.13. 'Recommendations for the Presentation of Plastics Design Data', British Standard 4618, British Standards Inst., London, 1970.

4.14. BERRY, D.B.S., 'Tests on full size plastics panel components', Chapter 9 in *The Use of Plastics for Load-bearing and Infil Panels,* (ed.) HOLLAWAY, L., Manning Rapley Publishing, 1976.

4.15. 'Methods of Testing Plastics (test methods 306D and 306E)', British Standard 2782, British Standards Inst., London, 1970.

4.16. A.S.T.M. D 1822-68, 'Tensile-Impact energy to break plastics and electrical insulating materials. Tests for', American Society for Testing and Materials, Philadelphia.

4.17. WESTOVER, R.F. & WARNER, W.C., 'Tensile impact test for plastics', *Mater. Res. Stands.,* 1, 11, November, 1961, pp. 867-871.

4.18. IVES, G.C., MEAD, J.A. & RILEY, M.M., *Handbook of Plastics Test Methods,* Iliffe, 1971.

4.19. NEUMAN, R.C., Paper VI, Proc. 21st Ann. Tech. Conf. S.P.E., March, 1965.

4.20. CAPRON, E., CROWDER, J.R. & SMITH, R.G., 'Appraisal of the weathering behaviour of plastics', B.R.E., CP 21/73, 1973.

4.21. SCOTT BADER, *Polyester Handbook,* Crystic Monograph No. 2.

4.22. BADER, M., BAILEY, J., & BELL, I., 'The effect of fibre-matrix interface strength on the impact and fracture properties of carbon-fibre-reinforced epoxy resin composites', *J. Phys., D : Appl. Phys.,* 6, 1973, pp. 572-586.

4.23. VINCENT, P.I., 'Short term strength and impact behaviour', in *Thermoplastics : Properties and Design,* (ed.) OGORKIEWICZ, R.M., Wiley, London, 1974.

4.24. Internal publication on cladding and panel units, The Agrement Board, Hemel Hempstead, Herts.

4.25. BONNIN, M.J., DUNN, C.M.R., TURNER, S., 'A comparison of torsional and flexural deformations in plastics', *Plastics and Polymers,* 37, December, 1969.

4.26. WILLIAMS, M.J., MOORE, D.R. & ELDRIDGE, J.W., 'An apparatus for accurate measurement of shear modulus in plastics', Unpublished paper, I.C.I. Plastics Division, Welwyn Garden City, Herts.

4.27. GRESZCZUK, L.B., 'Shear modulus determination of isotropic and composite materials', Composite Materials : Testing and Design ASTM STP 460, 1969.

4.28. GRESZCZUK, L.B., 'Douglas ring test for shear modulus determination of isotropic and composite materials', Proc. 23rd Ann. Tech. and Man. Conf. Soc. of the Plastics Industry, Washington D.C., February, 1968.

5 End-use performance properties of structural plastics and G.R.P. composites

5.1 Introduction

To a structural engineer the in-service and the miscellaneous properties of materials are just as important as the mechanical properties; indeed the long term in-service characteristics of a material are interrelated with its mechanical properties.

Durability is an aspect of performance which embraces most physical as well as mechanical and aesthetic properties and is therefore one of the most important qualities to be considered when plastics are used externally on buildings. It is obviously necessary that plastics show a degree of resistance to the degradative action of the weather and although most plastics are, to a certain degree, chemically inert, their surfaces and mechanical properties may change when exposed to the atmosphere, which is not desirable. Architects see the 'mellowing' of traditional materials as a gain in aesthetic appeal but wish plastics to retain their pristine condition to achieve a particular effect and do not regard any physical change as desirable. To the user, change in appearance is often the most significant form of deterioration, but aesthetic properties and limits of acceptability are the most difficult to define. These points will be discussed later in this chapter.

The chapter is divided into three parts. The first part has a section on the specification and quality control of civil engineering plastics. Thermal and chemical properties, sound insulation, light transmission and abrasion resistance are also discussed. The second part discusses the durability of G.R.P. in relation to its mechanical and physical properties. The final part attempts to present the various standards relevant to fire and to describe how improved fibre performance may be imparted to plastics structures through specific provisions as used in the construction industry.

PART I

5.2 Specification and quality control

The procedure for controlling a product's performance is known as quality assurance; the quality of the finished product will therefore be dependent upon this assurance. To be able to ensure that the components meet all the necessary requirements, it is essential to write a specification covering all aspects of performance and to ensure that the finished product meets this specification. The specification should include:

(a) raw material quality;
(b) an adequate design of all components;
(c) sufficient detail consistent with design requirements for the manufacture of the finished components.

Because there are several different manufacturing processes for glass reinforced plastics ranging from high output, high investment press moulding to low output, low investment hand lay-up, the quality control problems for the various processes are different. Once product, tool design and press operating techniques have been established, the hot press moulding automated stages involve fewer quality control problems compared to, say, the hand lay-up and spray-up techniques which are carried out at ambient conditions and in open moulds with little or no mechanization.

The quality control of any glass reinforced plastics product and the manufacturing processes to obtain it are the responsibility of many departments or sections within a firm; Dowson [5.1] has listed and commented on the roles of the various departments within a typical fabricating company and has also discussed the quality control related to materials and inclusions used in G.R.P.

The most important items for inclusion in the specification are mentioned below. The quality control for reinforced plastics mouldings is covered by the B.S. 4549 [5.2] and is based upon performance criteria; this control must be assessed by a routine programme of testing. Probably the most important part of B.S. 4549 is Part 1, which deals with mouldings such as chopped strand mat or randomly deposited glass fibre. This material is discussed in detail below as an example of these types of specifications.

(a) Materials
If the materials' specifications are to comply with a B.S. Specification, the suppliers must guarantee the material to this standard.

(b) Selection of test specimens and frequency of testing
Test specimens for the assessment of hardness and residue on ignition must be

taken from the actual composites. For other tests, specimens with identical fabrication histories are sufficient. An adequate number of samples must be selected to be representative of the product quality and appropriate tests undertaken.

(c) Preliminary examination

Each component must be examined visually for defects such as protruding fibres, pits, blisters, crazing and cracks, voids, bubbles, resin rich or resin starved areas, surface tackiness and presence of foreign matter.

(d) Dimensions

Tolerances on dimensions of the components must be agreed between architect and fabricator. Tolerances for the thickness of a composite are given in Table 5.1 (derived from Table 1 in B.S. 4549 Part 1).

Table 5.1   Tolerances for the thickness of a composite
(Values after B.S. 4549 Part 1, 1970, Table 1)

Nominal thickness (mm)	Open mould (mm)	Closed mould (mm)
Up to but not including 1.5	−0.25 to + 0.5	± 0.20
1.5 but not including 3.0	± 0.75	± 0.30
3.0 but not including 6.0	± 1.1	± 0.50
6.0 but not including 12.0	± 1.5	± 0.75
12.0 but not including 25.0	± 2.0	± 1.4
25.0 and over	± 3.0	± 1.9

(e) Glass content

If the nominal glass content is specified as $N\%$, the variation must be within $(N-2.5)\%$ to $(N+7.5)\%$. If, during the manufacture of a composite, no mineral fillers are used, the residue on ignition can serve as a quick control to the glass content. Alternatively, if mineral fillers are used, B.S. 2782 'Resin Content of Glass Reinforced Laminates' method 107K may be used as a basis for determining the glass content.

(f) Degree of cure

The degree of cure of a composite may be estimated by the Barcol hardness of the material. The minimum value should be at least 90% of the Barcol value specified by the resin manufacturer. The Barcol value is temperature dependent and therefore the test temperature with a tolerance of $\pm 2^\circ$ C on the minimum Barcol values for a range of temperatures must be specified. In certain cases it may also be necessary to determine the hardness of the gel coat.

It should be noted that the requirements of the B.S. 4549 Part 1 are not always satisfied, particularly in the open moulding process. Some fabricators have difficulty in satisfying the tolerances for the thickness of the composite and for the variation of the glass content.

5.3 Thermal properties

The thermal properties of glass reinforced plastics are important when used for structural components. Unreinforced or unfilled plastics have a very high coefficient of expansion which creates design and detailing problems when used in conjunction with traditional materials. Table 5.2 contains typical thermal properties of G.R.P. components and compares them with conventional engineering materials. One of the principal effects of incorporating glass fibres and fillers in the plastics materials is to reduce the coefficient of thermal expansion to a composite value of the same order as that for aluminium.

Table 5.2 Thermal properties of some engineering materials

Material	Coefficient of linear expansion ($°C \times 10^{-6}$)	Thermal conductivity (W/m $°C$)
Steel	11.3	46.0
Aluminium	23.0	140.0 – 190.0
Timber	5.4 – 54	0.124 – 0.24
Concrete	13.0	0.98
Glass fibre	8.6	1.02
Polyester resin	50.0 – 100.0	0.11 – 0.28
Glass fabric with polyester resin	9.0 – 11.0	0.3 – 0.35
Woven fabric with polyester resin	11.0 – 16.0	0.2 – 0.3
Hand lay-up C.S.M. laminate	22.0 – 36.0	0.2 – 0.24
S.M.C.	20.0 – 28.0	

The thermal conductivity of all plastics is low, as shown in Table 5.2; consequently, glass reinforced plastics of sufficient thickness are good insulators and when used with glass fibre wool or foamed plastics the composite construction has an extremely low U value. When reinforced plastics are used as the faces of an insulating sandwich construction, the resulting composite has a low thermal conductivity, and the likelihood of vapour condensing on its internal surface is small.

The temperature limit at which resins begin to lose their rigidity is known as the heat deflection temperature (H.D.T.). If the resin is reinforced this value rises by about $20°$ C. The heat deflection temperature represents a limiting factor in design as creep of the material under load becomes appreciable when approaching this temperature. The British Standard 4994 [5.4] does not permit the use of resins at temperatures greater than (H.D.T. – $20)°$ C.

British Standard Code of Practice 3, Chapter II [5.5] describes the factors affecting the internal environment of a building together with appropriate design recommendations. Thermal insulations of buildings are also controlled by the building regulations.

F

5.4 Chemical resistance properties

The chemical resistance of G.R.P. laminates used in the construction industry is basically a function of the quality of the gel coat as it is this coat which is exposed to the atmosphere. If specific chemical resistance is required, great care should be exercised in the choice of resins; chemical resistance is more sensitive than most other properties to the type of resin employed.

A convenient way of investigating the chemical resistance of various types of resins is to compare their performance under maximum working temperatures (M.W.T.) and in known chemical environments. Raymond [5.6] has given comparisons of some of Scott Bader's Crystic resins (the selection of crystic polyester resins given is of particular interest in the chemical plant market) in terms of the M.W.T. at which plant and vessels designed to B.S. 4994 will operate continuously for several years without serious deterioration. He has given the test procedures to obtain these results and has suggested that the onset of fracture could be due to permeation, solution of water soluble matter or swelling due to adsorption, all of which produce stress build up. Since osmosis plays an important part in the process of degradation, pure water and weaker salt solutions are generally more aggressive than the concentrated ones.

Raymond has also suggested resins which may be used at temperatures usually associated with civil engineering structures. These structures include G.R.P. storage tanks for drinking water, effluent pipes and chimneys. He has stated that it is often possible to tolerate a greater degree of deterioration of the composite when gases and fumes are involved. Fume stacks function perfectly well even when the surface is pitted beyond a state which for most liquid chemicals could not be tolerated; these M.W.T. would be determined by the structural limitations imposed by the heat deflection temperature.

5.5 Light transmission property

The light transmission of a translucent G.R.P. composite exposed to the atmosphere for a number of years is dependent upon the type of resin used and the resin content. Scott Bader [5.7] has observed a 14% reduction of light transmission over five years in their Crystic 191E resin reinforced with glass mat of fibre-matrix ratio 30–70% by weight and with a gel coat containing surface tissues. The same composite without a gel coat has a reduction of 30% and a resin glass mat with a fibre-matrix ratio of 35–65% has a reduction of about 70%; both these latter examples are over the same period of time as that of the former.

The accumulation of dirt on the external surface of G.R.P. laminates used as roof units is a further cause of reduction of light transmission and it is advisable to make provision for periodic cleaning of such surfaces. Common problems with fire retardant resin laminates are discoloration, variations in the hue, and the appearance of dirt particles on the surface of translucent laminates.

5.6 Sound insulation

In all buildings sound transmission must be reduced to a minimum. The ideal way of achieving this is to incorporate heavy bulk materials in the construction. As with many other materials, reinforced plastics are not normally used in bulk and consequently sound reduction is difficult to achieve. The required degree of sound insulation may be provided by utilizing a composite design. In a sandwich construction this can be achieved by a filled foam core or by using a plaster board as a lining material in which the sound is reflected or absorbed.

Expanded polystyrene is sometimes used as ceiling tiles but these do not absorb sound by porosity as do most acoustic tiles. They can however give a measure of absorption when mounted on battens with an air space between the tiles and the backing surface.

British Standard Code of Practice 3, Chapter III[5.8] and British Standard Code of Practice 153, Part 3 [5.9] give guidance on the design of building elements and components to ensure adequate sound insulation and noise reduction for various types of buildings and locations.

5.7 Abrasion resistance

To obtain an acceptable abrasion resistance which depends upon hardness and toughness, it is essential to use a gel coat with surface tissues. Table 5.3 gives typical abrasion resistance values for 'good' and 'poor' gel coats.

Table 5.3 Abrasion resistance of G.R.P. laminates
(By kind permission of Fibreglass Ltd., Reinforcements Division)

Material	Surface roughness (μm)		Wear (μm)
	as received	after 1200 cycles test with abrasive cleaner	
Good gel coat on laminate	0.25 – 0.76	1.0	64
Poor gel coat on laminate	0.25 – 0.76	5.3	192

PART II

5.8 Durability

The term *durability* is used to denote the period of time over which a material will perform its allotted task in its given environment. The specification of the durability should state that the components must conform to the performance specification for the expected life of the building; however, for some components a shorter life may be acceptable. For this to be realistic the appearance specification will need to incorporate allowances for any anticipated changes.

Durability is often difficult to assess and requires a keen judgement of what constitutes sufficient duration and adequate performance. One year's duration may be adequate when considering a chemical environment and Yovino and Dunwoody [5.10] have rated the resistance of G.R.P. and other materials to aggressive fluids over this period according to performance. On the other hand the useful life of a composite exposed to the weather as part of a building may be as long as 50 years. This poses a problem when attempting to improve formulations since the results of field performance over limited periods have to be extrapolated to the specified length of time.

Adequate performance is also difficult to assess from a durability point of view because the aesthetic properties that are difficult to quantify may be important. An unacceptable condition for roof lights may be precisely defined in terms of light transmission. In the case of opaque cladding, uneven discoloration and grime accumulation may manifest the changes that are unacceptable but are difficult to define.

Appearance of the exposed surfaces of a plastics component may change significantly on weathering, and this in some cases may be sufficient to render that article aesthetically unacceptable. Mechanical properties of plastics may also decline on weathering, especially tensile strength and impact strength which are sensitive to surface deteriorations. Recently, the Building Research Establishment has initiated a research programme to investigate weathering of G.R.P. under stress (see section 5.8.1).

Crowder [5.11] has stated that changes in optical, mechanical and surface appearance properties are not interrelated. Consequently, observations of a simple property are not an adequate measure of the weathering performance of a plastics material. The phenomena which lead to failure of plastics due to weathering are complex and vary with different materials.

The most significant single factor responsible for the breakdown of plastics and for producing changes in colour is sunlight, and particularly its ultra violet component. Other factors influencing this degradation are temperature, humidity and the permeability of the exposed surface; the last controls the penetration of oxygen and the escape of the products of degradation.

Fillers and pigments have a major effect on appearance and durability of plastics. The use of flame retardant additives may aggravate yellowing of the exposed surfaces of the plastics.

Two important factors upon which the durability and weathering performance of a G.R.P. composite depend are:

(a) the choice of resins for the gel coat and for the matrix of the composite; the appropriate type of reinforcement must also be used;
(b) the provision of the required level of quality control to ensure a suitable production environment, a correct fabrication procedure and an adequate cure for the resin.

Although it is clear that the first factor has a considerable influence on the performance of G.R.P., it is difficult to give any quantitative information as resins made to present day specifications have not undergone long term weathering and durability trials to a time scale comparable to the assumed life of a building construction (usually 50 years). The environment in which the proposed building panel is to perform has a major effect on the durability of G.R.P. The extent to which the mechanical properties of the material decline with time due to the natural agents has not been fully established (see section 5.8.1); however, the changes in physical appearance of G.R.P. panels in service have been observed. Resin producers have stated that these changes are not associated with a reduction in the mechanical properties of G.R.P. but are merely based upon surface phenomena.

Accelerated tests may be undertaken to provide information on the long term behaviour of plastics but inevitably they have limitations. This is principally due to:

(a) the difficulty of correlating the results of the accelerated laboratory tests with normal weathering conditions;
(b) the lack of an interrelationship between the optical, mechanical and surface appearance properties of plastics materials.

The British Standard 4618 [5.12] recommends natural weathering trials to be made with exposure times of 3 months, 6 months, 1, 2, 4, 6, 8 and 10 years and such longer periods as are necessary. Climates are also broadly classified into five different types:

(a) hot and wet;
(b) hot and dry;
(c) mesothermal;
(d) temperate;
(e) cold.

It is recommended that trial conditions should simulate those of normal service; in addition certain other tests should also be performed. For instance, the effects of periodical cleaning should be observed on a second series of specimens.

During weathering tests the specimens under examination are assessed against control specimens stored under defined conditions, for any or all of the following:

(a) dimensions;
(b) visual appearance;
(c) mechanical properties;
(d) biological attack.

5.8.1 The weathering of G.R.P. under stress

Over a number of years the B.R.E.[5.13] has studied the long term weathering behaviour of plastics but generally these exposure trials did not subject the specimens to external stress. Other investigators have studied the long term behaviour of unstressed G.R.P. composites in different situations which have included immersion in water, chemical and thermal environments [5.14] [5.15] [5.16]. Some studies have been made of the combined effects of weathering and stress on G.R.P. but these have tended to rely upon the creep characteristics of the material rather than its residual strength [5.17].

The B.R.E. is currently carrying out a study of the long term behaviour of G.R.P. composites when exposed to the weather and under an applied tensile load; the changes in the specimens' properties are determined by tensile and flexural tests. Norris *et al.* [5.18] and Probert *et al.* [5.19] have given the test results of the first and second years respectively. The two specimen types used in the investigation are manufactured from materials typical of those used in the construction industry; specifically, these are:

(a) a flame retardant chopped strand mat with a gel coat of isophthalic resin without flame retardant;
(b) a flame retardant chopped strand mat with a gel coat incorporating a flame retardant.

Both are capable of achieving a Class II spread of flame rating (see section 5.9.1.1). Specific test procedures for the investigations have been laid down and the results found must be considered in the light of these conditions.

The loaded specimens were inspected at regular intervals and during the first 2½ year period no apparent failure of the specimen or the gel coats was noted. However, specimens incorporating the flame retardant gel coat have yellowed more than those without the flame retardant in the gel coat.

After this time span the research team have suggested that the changes in most properties are of a relatively small order. The main results so far are that the specimens containing the flame retardant gel coat had surface degradation which produced significant reductions in the stress levels so as to produce surface crazing, but this effect did not occur with the general purpose gel coat specimens. Over the two year period of exposure, the effect of these changes of stress, for an applied stress of up to 25% of short term ultimate, was found to be of negligible significance. The modulus of elasticity in tension increased in value for all exposed specimens, but there was a decline in the flexural modulus of the order of 15%. The research workers have suggested that the effect of U.V. radiation in increasing post-cure and the resulting hardening from increased cross-linking may explain the increase in tensile elasticity; no definite reason can be given for the fall off in the value of flexural modulus.

PART III

5.9 The fire behaviour of plastics

One of the chief concerns of the architect and engineer using plastics in the construction industry is the problem associated with fire. In this context it is most important for the reader to realize that the fire problems here discussed refer to G.R.P. and not to plastics generally. It is still common within the civil engineering profession to hear references made to the fact that 'plastics' have a high fire risk. 'Plastics' cannot be treated generally and it will be shown that G.R.P. has no greater fire risk than many other civil engineering materials.

A brief description of the more relevant fire tests laid down in the British Standards will be given before the remedial measures for G.R.P. are discussed.

Because plastics materials are organic in nature and are inherently combustible, they will decompose or burn when in contact with fire. The steady growth of plastics consumption in the construction industry [5.20] has necessitated a fuller understanding of the potential fire hazards and fire behaviour, particularly of the more commonly used plastics.

A fire usually develops through three stages:

(a) fire development stage which includes ignition, flammability, spread of flame, heat release and production of smoke and toxic gases;
(b) the steady combustion stage in which the fire continues in a steady regime;
(c) the fire decay stage in which all the available combustible materials have been exhausted. Generally this stage is not reached.

Bub [5.21] has expounded the German stages of fire progression.

The behaviour of materials in the first stage is termed reaction to fire. The fire resistance of a building element is a measure of its ability to confine the fire within a compartment whilst the unaffected part of the structure remains stable throughout the time of the fire. The fire resisting requirements of different elements of a building are therefore related to the amount of combustible material present in the compartment, its function and its position relative to building boundary lines.

The Building Regulations [5.22] require that, depending upon their use, building components or structures should conform to given standards of fire safety. The fire tests by which these are measured fall into two categories:

(a) reaction to fire—tests on materials;
(b) fire resistance—tests on structures.

The tests under these two headings are laid down in BS 476: Part 4, 5, 6 and 7 and BS 476: Part 3 and 8 respectively [5.23], (Part 3 covers tests on roofs). A glossary including terms used in BS 476 has been established and is published in BS 4422 [5.24].

5.9.1 Reaction to fire

Plastics are classified as combustible materials when tested to BS 476 : Part 4, 1970. Materials used as linings to walls and ceilings are mostly required to achieve a Class 1 rating on BS 476 : Part 7, although relaxation to Class 3 for certain domestic areas is permitted. G.R.P. may also be used in roof lights as part of a surface of a ceiling, in limited areas and spans under the Building Regulations Section E15 [5.22] giving relaxation to Class 3 in accordance with BS 476 : Part 7. Where a higher degree of safety is required (e.g. on escape routes) Class 0 is specified by the Building Regulations, based on a performance Index $I \vartriangleright 12$ and $i_1 \vartriangleright 6$ (see section 5.9.1.2 for explanation of index of performance) when tested to BS 476 : Part 6. For cladding materials within 1 m from the boundary or on upper parts of a high building (above 15 m from the ground) Class 0 is also specified, but on the lower parts of this latter building, the specification requires timber of not less than 9 mm finished thickness or the maximum performance Index of 20, when tested to BS 476 : Part 6. By using inorganic fillers in plastics, various performance levels may be obtained. However, a very low index can only be achieved if the amount of inorganic material present in the specimen is increased substantially [5.25].

5.9.1.1 The surface spread of flame test
BS 476 : Part 7 [5.23] specifies a vertically mounted radiation panel of 900 mm × 900 mm. A test sample of the material shall comprise six specimens of surface area 230 mm × 900 mm and thickness equal to the composite under consideration where this does not exceed 50 mm. If it is thicker, the material should be reduced to 50 mm. These specimens are mounted vertically with the longitudinal axis horizontal. The radiation received by the specimen will vary from 37 kW/m² , at the end nearest to the heat source, to 7.5 kW/m² at the free end. The hotter end is then ignited for one minute.

Table 5.4 Classification for the spread of flame test

Spread of flame at 1½ min	Spread of flame at 10 min	Classification
Not more than 165 mm Tolerance on one specimen 25 mm	Not more than 165 mm Tolerance on one specimen 25 mm	Class 1
Not more than 215 mm Tolerance on one specimen 25 mm	Not more than 455 mm Tolerance on one specimen 25 mm	Class 2
Not more than 265 mm Tolerance on one specimen 25 mm	Not more than 710 mm Tolerance on one specimen 25 mm	Class 3
Exceeding Class 3 limits		Class 4

The spread of flame on all specimens of the sample is measured and should fall into one of the categories laid down in Table 5.4. The surface spread of flame test has limitations, although these can be accommodated. It is an open type test and does not provide direct information on the effects of ignitability, rate of heat release or smoke production in different orientations of the specimen. However, it is extensively used in conjunction with BS 476 : Part 6.

It is possible to achieve a Class 1 spread of flame with plastics materials but it is not generally recommended due to the poor weathering performance associated with highly filled plastics [5.26].

5.9.1.2 The fire propagation test

In BS 476: Part 6 a test specimen of dimensions 228 mm × 228 mm and nominal thickness not greater than 50 mm is mounted on a non-combustible backing and the whole forms the lining of one vertical side of a non-combustible cubical furnace. Initially, the specimen is subjected to a low heat intensity of 5.0 kW/m² and after 2¾ minutes the intensity is raised to 50 kW/m² for the 20 minute test duration. The emitted gases are directed through a chimney and cowl arrangement and the time-temperature curve for these gases is plotted at various intervals up to the maximum time of 20 minutes. The results are compared with those obtained from a known non-combustible material. An 'index of performance' I is calculated using the following expression [5.23]:

$$I = \sum_{½}^{3} \frac{\theta_m - \theta_c}{10t} + \sum_{4}^{10} \frac{\theta_m - \theta_c}{10t} + \sum_{12}^{20} \frac{\theta_m - \theta_c}{10t}$$

$$I = i_1 + i_2 + i_3$$

where i_1, i_2 and i_3 are sub-indices for the three periods measured at ½, 1 and 2 minute intervals respectively.

θ_m is the temperature rise in deg C recorded for the material at time t

θ_c is the temperature rise in deg C recorded for the non combustible material at time t

t is the time in minutes from the beginning of the test.

The sub-index i_1 is a measure of early heat release under the influence of gas flames alone. It has been shown in reference [5.25] that the sub-index i_3 has a relatively low value compared with i_2. Materials which have a high potential of heat release after prolonged exposure to severe heating conditions show a relatively high i_3, unlike other materials which ignite at lower intensities. A low performance index represents a small contribution to the growth of fire.

Table 5.5   Performance indices for G.R.P. and part plastic composites (Derived from ref. [5.25])

Material	Treatment	Thickness	I	i_1	$i_2 + i_3$
G.R.P. (C.S.M.) with a gel coat		3.6	22.9	9.7	11.0
G.R.P. (W.R.) with a gel coat		4.6	18.9	9.0	9.3
G.R.P. (C.S.M.) sheet	with flame retardant additives	3.0	11.1	4.0	6.2
G.R.P. (W.R.)	intumescent gel coat	4.9	9.1	1.1	7.1
G.R.P. (C.S.M.)	flame retardant grade	3	11.2	4.0	7.2
Polyurethane foam	flame retardant grade	35	38.7	28.3	11.4
Polyurethane foam	flame retardant grade	13	28.6	23.4	5.2
Acrylic sheet	flame retardant grade	3	39.8	20.0	19.8
PVC		3	16.8	5.9	10.9
Melamine phenolic laminate	flame retardant grade	2	7.2	1.5	50
Steel sheet	PVC coat 0.3 mm	3	5.5	2.2	3.3
Plaster board		9	9.7	5.7	4.0
Plaster board	emulsion painted	13	9.0	5.2	3.8
Plaster board	PVC facing 0.2 mm	9	10.0	5.4	4.6

In the Building Regulations [5.22] the Class 0 fire rating is used to define materials which, as well as having a performance index of not more than twelve, comply with the Class 1 surface spread of flame to the BS 476 : Part 7. The Class A fire rating is used to define materials which have a Class 1 surface spread of flame but fail to achieve an index of performance below 12. A performance index of below 20 indicates acceptability for use up to a height of 15 m on the external walls of buildings which are over 15 m high.

The fire propagation test is probably more relevant to G.R.P. composites than is the spread of flame test, due to the potential heat contribution. Table 5.5 contains some performance indices for various G.R.P. laminates and conventional materials and it may be observed that flame retardant grade G.R.P. laminates are able to achieve indices comparable to those for plaster board.

5.9.1.3 Smoke generation characteristics
In addition to assessing the combustion and flame spread properties of plastics, it is also necessary to investigate their smoke generation characteristics; smoke and evolved gases can create both physiological hazards, due to toxicity and irritancy effects, and pyschological hazards, due to reduced visibility through smoke and state of panic.

Experimental procedures for the assessment of toxicity and irritancy of smoke are not well advanced and because of this a Draft for Development [5.27] was prepared with the intention of issuing it at a later date as Part 9 of BS 476 (Fire Tests on Building Materials and Structures). The Draft for Development was rejected because of lack of reproducibility and the N.B.S. test (National Bureau of Standards, Washington, U.S.A. test) was proposed in its place. The ISO (International Standards Organisation) test is still not entirely complete but is similar in concept to the N.B.S. test. Together with the proposed BS 5111 [5.28], which is a laboratory test procedure for the determination of smoke generation characteristics of cellular plastics and cellular rubber materials, a greater awareness for the design of smoke emission should result.

5.9.1.4 Optical density of smoke
The Draft for Development 36 (D.D.36)[5.27] has assessed the optical density of smoke produced by various materials. Provided the vision has not been affected by any irritancy, the optical density will be related directly to the visibility through the smoke.

Factors influencing the production of smoke from a given material are:

(a) the mode of combustion, i.e. whether the material is burning with a flame or is smouldering;

(b) the state of ventilation;

(c) the intensity of incident radiation.

The above draft has described the measurement of optical density of smoke produced from small specimens in standard conditions, under both flaming and non-flaming modes of combustion. During the experiment the smoke produced in the fire propagation apparatus is evolved into a smoke-tight chamber where it is diluted. The optical density is measured in terms of the transmittance of a parallel beam of light falling on to a photoelectric cell and is calculated using the following expression:

$$D = -\log_{10} \frac{\bar{T}}{100}$$

where $\bar{T}$ is the mean transmittance of three specimens for each material under given combustion conditions. The calculation of $\bar{T}$ must conform to the method laid down in D.D.36.

The greater the amount of smoke produced from a specimen, the higher will be the optical density or the lower will be the visibility. Although no performance requirements in terms of the above test have yet been proposed, the limited data [5.29] available suggest that under non-flaming conditions of combustion, mineral-based products with relatively insignificant quantities of combustible materials have a low optical density, whereas G.R.P. laminates and hardboard have a comparatively high one.

5.9.2 Fire resistance

In the BS 476 Parts 3 and 8 [5.23] and most other national Standards, fire resistance applies only to structures and not to the individual materials of which the elements are made. It is generally measured in length of time (e.g. ½ hour, 1 hour, etc.).

The fire resistance tests enable the assessment of the ability of building components to retain their structural stability and to resist the passage of flame or hot gases.

5.9.2.1 Test methods and criteria for the fire resistance of elements of building construction

The test component should be representative of the element of construction in terms of fixings, finishes and manufacture, and the component should include at least one of each type of joint. Before the heating test, a load bearing component is subjected to the equivalent working stresses for the complete structure and this load is maintained during the test period; the standard time for the test is 6 hours at a temperature of 1200° C. During the test, observations are made on:

Stability: the deformation of the specimen, any collapse or other factors which could affect its stability are noted. For non-load bearing construction, failure is assumed to occur when collapse of the specimen takes place. Load bearing structures should support the test load during the prescribed heating period and for a further 24 hours.

Integrity: the presence of cracks or other openings. Failure is assumed to occur if flaming or hot gases pass through these openings and is determined by holding a cotton wool pad close to the aperture in the component at frequent intervals and for not less than 10 seconds, in order to observe whether the hot gases cause it to ignite.

Insulation: the temperature condition of the unexposed face of the specimen by continuous recording.

No published work on the testing of G.R.P. components to BS 476: Part 8, is available, although the Fire Research Station has undertaken tests on G.R.P. components for commercial organizations; the information derived from these tests has not, however, been for general publication.

5.9.2.2 External fire exposure roof test

BS 476: Part 3 1975 [5.23] specifies that two successive tests shall be performed: the preliminary ignition test and the fire penetration and surface ignition test. In the former the specimen is exposed to a flame for one minute and on removal observations are made for any continued flaming on the upper surface or fire penetration to the underside. If penetration does occur no further test is undertaken.

In the fire penetration and surface ignition test three specimens are tested and radiant heat of $14.6 \pm 0.5 \, kW/m^2$ is applied to the upper surface of each for 60 minutes and may be extended to a maximum of 90 minutes unless before this time penetration has occurred. After the specimen has been exposed to the radiant heat for five minutes, the test flame is moved slowly over the surface at varying intervals and for 1 minute duration. The time and occurrence of any flaming on the upper and lower surfaces and the development of holes and fissures are noted. Visual observations of changes in appearance, flaming and dripping are made and the minimum distance of the lateral flame spread on the upper surface is measured. The results are expressed in letter notation as follows:

The letter X indicates that:
(a) the duration of flaming exceeds 5 minutes after the withdrawal of the test flame, or
(b) the maximum distance of flaming in any direction is greater than 370 mm.

The letter P indicates that:
(a) the duration of flaming is less than 5 minutes, or
(b) the maximum distance of flaming in any direction is less than 370 mm.

The extent of surface ignition shall be given at 60 minutes or at the time of penetration to the nearest 25 mm. The classification P60, therefore, indicates that the specimen passed the preliminary test and that fire penetration did not occur within one hour.

Although the above classification has been introduced as a British Standard, it has not yet been included in the Building Regulations and the earlier edition of the British Standard is still active and will probably remain so for some time. Consequently, the notation for classification of the earlier Standard will be given here.

A prefix Ext S or Ext F is used to denote the flat or inclined test respectively, followed by two letters denoting the results of the penetration test (given in Table 5.6) and the spread of flame test (given in Table 5.7). The final letter indicates the extent of dripping on the underside of the specimen. The performance of G.R.P. in structural roofing applications should in general be at least Ext SAA or Ext FAA.

Table 5.6 Notation for results of penetration test

Penetration time	Letter code
More than 1 hour	A
More than ½ hour	B
Less than ½ hour	C
Fails preliminary test	D

Table 5.7 Notation for results of spread of flame test

Spread of flame	Letter code
None	A
Less than 534 mm	B
More than 534 mm	C
More than 381 mm in preliminary test	D

5.9.3 Methods of imparting flame retardancy to G.R.P.

A coating of intumescent resin retards combustion of the organic materials present in the main structure of the G.R.P. laminate and also significantly reduces the area affected by the flame, drastically reducing the volume of smoke and practically eliminating burning. This resin is specifically formulated and ready to apply by brush, after the addition of a catalyst, to a G.R.P. laminate. It gives fire protection by foaming in situations where G.R.P. laminates are exposed to direct flaming. This carbonaceous foam and the inert gases produced insulate for a period of time, dependent upon the severity of the heating conditions, the main structure of the G.R.P. laminate against flame. It is usually applied to the inside surface of a laminate. A flame retardant laminate with a gel coat especially formulated for specific weathering properties cannot usually achieve a Class 0 spread of flame and is generally limited to a Class 2.

There is a serious potential smoke problem in the fire retardant grades of polyester resins and therefore in future, with the possible introduction of a British Standard based upon certain relevant and equivalent sections of the Draft for Development 36 and the N.B.S. test, the following structural applications may be affected:

(a) External applications where all G.R.P. roofings and external elevations are used, especially areas adjacent to the fire exits and windows of buildings.
(b) Internal applications where G.R.P. is used for space dividers, lighting and decorative panels and the underside of G.R.P. roofing panels, particularly in escape routes.

(The E.E.C. requirements under the proposals of the working party DG XI [elimination of technical barriers to trade in building materials—reaction to fire of building materials] have suggested the adoption of the N.B.S. test; if accepted, it is likely that it will be introduced into Britain. Until this happens the I.S.O. test is available.)

Mineral fillers such as calcium carbonate, china clay, asbestos and alumina trihydrate impart varying degrees of flame retardancy to G.R.P. These fillers generally reduce the amount of combustible material by replacing some of the resin in the body of a laminate. In the case of alumina trihydrate the filler has additional effects of flame retardancy and smoke reduction due to the significant quantity of heat absorbed by the endothermic reaction during the decomposition of this material into alumina and water [5.30].

To achieve any significant level of flame retardancy, filler contents higher than 100 parts of filler to 100 parts of resin by weight are required. In the hand lay-up method of G.R.P. production the filler content is restricted to an upper limit of 75 parts per weight of resin, and is generally about 45. It is, therefore, necessary to add antimony trioxide and chlorinated paraffin in the ratio of between 2 to 15 and between 1 to 10 parts per weight of resin respectively, depending upon whether Class 2 or Class 1 is desired.

Another method of providing flame retardancy in G.R.P. is to sandwich stone chippings between two thin films of gel coat in a normal hand lay-up or spray-up technique; it is necessary to allow the two films of the gel coat to gel before applying the subsequent layers.

Fire penetration resistance can be achieved by the use of a cheap fire-resistant inner lining such as asbestos or plasterboard. Alternatively a sandwich laminate can be used with a fire resistant core material; one hour fire ratings have been achieved by this method.

5.9.4 Full scale fire tests

There are certain members within the civil engineering profession who suggest that testing small coupons is not a realistic means of determining the fire resistance or reaction to fire of plastics materials: they argue that the only way to determine these properties is to undertake a full scale test. A number of these tests have been performed and although it is outside the scope of this book to discuss them, references [5.31] and [5.32] give details of two such tests.

It must be said, however, that most engineers accept that the tests laid down in BS 476 have their limitations, but by establishing points of comparison they enable sensible assessments to be made of the overall risk and possible fire behaviour. Equally *ad hoc* tests have their limitations. Rogowski [5.33] has stated that the standards of fire performance are based on standard fire tests applicable to materials or constructions. No scale tests can predict fully the performance of a material when incorporated in a structure.

References

5.1. DOWSON, B., 'Quality control limitations in reinforced plastics manufacture', Conference on Reinforced Plastics in Building and Construction, The Plastics and Rubber Institute, April, 1976.

5.2. 'Guide to quality control requirements for reinforced plastics mouldings', British Standard 4549 (Part 1 : Polyester resin mouldings reinforced with chopped strand mat or randomly deposited glass fibres), British Standards Inst., London, 1970.

5.3. 'Methods of testing plastics', British Standard 2782. Part 3, British Standard Inst., London, 1965.

5.4 'Vessels and tanks in reinforced plastics', British Standard 4994, British Standards Inst., London, 1973.

5.5. 'Thermal insulation in relation to the control of environment', British Standard Code of Practice 3, Chapter II, British Standard Inst., London, 1970.

5.6. RAYMOND, J., 'Reinforced plastics in chemical environments', Paper 9, Reinforced Plastics : Design and Technology, N.E.L., November, 1975.

5.7. SCOTT BADER, *Polyester Handbook,* Crystic Monograph No. 2.

5.8. 'Sound insulation and noise reduction', British Standard Code of Practice 3, Chapter III, British Standards Inst., London, 1972.

5.9. 'Sound insulation', British Standard Code of Practice 153, Part 3, British Standards Inst., London, 1972.

5.10. YOVINO, J. & DUNWOODY, G.T., Paper 18, Symposium on Reinforced Plastics and Composites, Sydney, 1970.

5.11. CROWDER, J., Building Research Station, Garston, Herts., private communication.

5.12. 'Recommendations for the presentation of plastics design data', British Standard 4618 (Part 4: Environmental and chemical effects), British Standards Inst., London, 1972.

5.13. CROWDER, J.R., B.R.E. Miscellaneous Papers 2, July, 1964.

5.14. RAWE, A.W., 'Environmental behaviour of glass-fibre reinforced plastics', *Trans. J. Plast. Inst.,* **30**, February, 1962.

5.15. OSWITCH, S., 'Some aspects of the thermal and chemical ageing of polyester/glass laminates', *Reinforced Plastics,* **8**, December, 1963.

5.16. GOLDFEIN, S., 'Prediction of mechanical behaviour of plastics undergoing decomposition from the combined effects of environmental exposure of stress', *J. Appl. Pol. Sci.,* **10**, 1966.

5.17. ZILVAR, V., 'Das Kriechverhalten von glasfaserverstärkten polyesterharzlaminaten bei Freibewitterung', *Kunststoffe,* **59,** 12, 1969.

5.18. NORRIS, J.F., CROWDER, J.R. & PROBERT, C., 'Weathering of glass-reinforced polyesters under stress—short term behaviour', *Composites,* **7,** 3, July, 1976.

5.19. PROBERT, C., NORRIS, J.F. & CROWDER, J.R., 'Weathering of plastics under stress', Symposium: The weathering of plastics and rubbers, Paper D3, The Plastics and Rubber Institute, June, 1976.

5.20. Fire Research Station and Building Regulations Professional Division 'Polymeric materials in fire', B.R.E., CP. 91/74, 1974.

5.21. BUB, H., 'Problems of fire safety when using plastics materials'. Chapter 12 in *The Use of Plastics for Load Bearing and Infil Panels,* (ed.) HOLLAWAY, L., Manning Rapley Publishing, 1976.

5.22. Part E. Structural Fire Precautions of the Building Regulations, 1976, H.M.S.O.

5.23 'External fire exposure roof test', BS 476, Part 3, 1975; 'Non-combustibility test for materials', BS 476, Part 4, 1970; 'Ignitability test for materials', BS 476, Part 5, 1968; 'Fire propagation test for materials', BS 476, Part 6, 1968; 'Surface spread of flame tests for materials', BS 476, Part 7, 1971; 'Test methods and criteria for the fire resistance of elements of building construction', BS 476, Part 8, 1972; British Standards Inst., London.

5.24 'Glossary of terms associated with fire', BS 4422, Parts 1,2,3,4,5, British Standards Inst., London 1969–1976.

5.25. ROGOWSKI, BARBARA F.W., 'The fire propagation test; its development and application', F.R.T.P., No. 25, 1970, H.M.S.O.

5.26. *Building Research Establishment Digest,* 161, H.M.S.O., January, 1974.

5.27. Draft for Development 36, 1974, British Standard Inst., London.

5.28 'Laboratory methods of test for determination of smoke generation characteristics of cellular plastics and cellular rubber materials', Part 1, BS 5111, 1974, British Standard Inst., London.

5.29. BOWES, P.C. & FIELD, P., Fire Research Note No. 749, 1969.

5.30. Alcoa Product Data, November, 1972.

5.31. Report on Special Investigation made on behalf of Lancashire County Council, County Architect's Department, County Hall, Preston, Lancashire, by Warrington Research Centre, Warrington WA2 7JE, Report No. W.R.C.S.I., No. 11006, 1974.

5.32. 'Report on experiments and reaction to fires of Mondial House Cladding', in *Glass Reinforced Resins for Structures,* Peter Hodge and Associates, Crewkerne, Somerset.

5.33. ROGOWSKI, BARBARA F.W. 'Plastics in buildings—fire problems and control', Building Research Establishment Current Paper CP 39/76, June, 1976.

Bibliography of standards concerned with the quality control of reinforced plastics

5.34. 'Unsaturated polyester resin systems for low pressure fibre reinforced plastics', BS 3532, British Standard Inst., London, 1962.

5.35. 'E Glass fibre chopped strand mat for the reinforcement of polyester resin systems', BS 3496, British Standards Inst., London, 1973.

5.36. 'Quality mark scheme', British Plastics Federation, Reinforced Plastics Group.

5.37. 'Common directive for the assessment of products in G.R.P. for use in building', Agrement Board, MOAT 9, 1973.

G

Appendix 5.1

The approximately equivalent B.S.S., A.S.T.M. and I.S.O. codes are given below:

B.S.S.	A.S.T.M.	I.S.O.
BS 476: Part 3 'External fire exposure roof test'	E 108–75 'Standard method of fire tests of roof coverings'	T.C. 92 WG4 'Fire tests on building materials and structures'
BS 476: Part 4 'Non-combustibility test for materials'	E 136–73 'Standard test methods for non-combustibility of elementary materials'	R. 1182 'Non-combustibility test for building materials' 1st ed., Jan 1970
BS 476: Part 5 'Ignitability test for materials'	D 1929–68 (1975) 'Ignition properties of plastics, Test for'	
BS 476: Part 6 'Fire propagation test for materials'	No equivalent	
BS 476: Part 7 'Surface spread of flame test for materials'	E 84–76 'Standard test method for surface burning characteristics of building materials' E 162–76 'Standard test method for surface flammability of materials using a radiant heat energy source'	R 834 'Fire resistance tests—elements of building construction' 1st ed., 1975
BS 476: Part 8 'Test methods and criteria for the fire resistance of elements of building construction'	E 119–76 'Standard methods of fire tests of building construction and materials'	

Appendix 5.1 (continued)

B.S.S.	A.S.T.M.	I.S.O.
BS 4422 'Glossary of terms associated with fire'	E 176-73 'Standard methods of terms relating to fire tests of building construction and materials'	I.S.O. 3261 Fire tests—vocabulary 1st ed., 1975
BS 4735 'Laboratory methods of test for assessment of the horizontal burning characteristics of specimens no larger than 150 mm × 50 mm × 13 mm (nominal) of cellular plastics and cellular rubber materials when subjected to a small flame	D. 1692-76 'Rate of burning or extent and time of burning or both of cellular plastics using a supported specimen by a horizontal screen, Test for'	I.S.O. T.C. 45 Rubber and Rubber Products I.S.O. T.C. 61 Plastics of the I.S.O.
BS 2782 'Methods of testing plastics'	A.S.T.M. Standards on Plastics Parts 34, 35, 36 Part 34: Plastics pipes— Part 35: Plastics general method of testing nomenclature paints and applied coatings Part 36: Plastics specifications: methods of testing pipes, film, reinforced and cellular plastics; fibre composites	I.S.O. T.C. 138/WG6 Proposed standard A.S.T.M. specification for reinforced plastics mortar, low head pressure pipes At the end of B.S. 2782 1970 a comparison is made with similar methods standardized or in the course of standardization by T.C. 61 of the I.S.O. In more recently published sections of B.S. 2782, and if applicable, the introduction to each part gives the methods of I.S.O. which are in general agreement with the B.S.S.

Appendix 5.1 (continued)

B.S.S.	A.S.T.M.	I.S.O
BS 4994 'Vessels and tanks in reinforced plastics'	D 3299–74 'Filament wound glass fibre reinforced polyester chemical resistant tanks, Specification for'	
BS 4549 'Guide to quality control requirements for reinforced plastics mouldings'	No equivalent	
BS 3532 'Unsaturated polyester resin systems for low pressure fibre reinforced plastics'	No equivalent	
BS 4618 'Recommendations for the presentation of plastics design data'	No equivalent	

6 Low density rigid foam materials, sandwich construction and design methods

6.1 Introduction

In previous chapters the properties and fabrication techniques of a single skin composite of G.R.P., capable of being used for small components as well as large folded plate structures, have been discussed. If it is necessary to increase the stiffness of the overall structure or the individual laminates, beyond the value for the material, a sandwich construction consisting of the reinforced polyester composites bonded to a low density core may be used. The face then would generally support the bending moments and axial forces within the composite and the core would take the majority of the shear. The cores may be of low density plastics and the three main types of foam which are used in the construction industry and which will be considered in this chapter are:

  (a) rigid polyurethane;
  (b) phenolic;
  (c) polyvinylchloride.

The first two plastics are thermosetting resins and the last one is a thermoplastic resin.

After the above three foams have been discussed, some design methods for sandwich construction will be given. Within one chapter it is impossible to deal fully with the design methods of sandwich construction. The chopped strand mat is the laminate used mainly in the construction industry as the face material of sandwich construction; consequently the design methods given here are generally applicable only to isotropic faces and isotropic core materials.

6.2 Rigid plastics foams

Rigid foams are two phase systems of a gas dispersed in a solid plastics; in most cases the solid plastics represents only a minor proportion of the total volume but contributes largely to the properties and utility of the foam. They are produced by adding a blowing agent to chemical formulations and this causes the materials to expand and to increase their original volume many times by the formation of small cells. Like solid plastics, rigid plastics foams can be thermoplastics or thermosetting plastics materials, and they share the advantages and limitations of the solid phase; in addition, the density, cell geometry and gas phase composition can be varied to modify their products.

6.2.1 Foam structure

The gas in a plastics foam is distributed in voids referred to as 'cells', whilst the solid plastics enclose these voids to form the cell walls. A foam in which the cells are discrete or disconnected units and whose gas phase is not continuous is said to have a 'closed cell' structure and is essentially air and vapour tight, whilst a foam in which the cells are interconnected and whose gas is continuous is said to have an 'open cell' structure. With this latter foam, free movement of air and vapour through the volume of material is permitted. Most rigid plastics foams, however, are neither completely open nor completely closed celled but are characterized by a 'fraction' of open or closed cells.

The arrangement of the gas and solid in a plastics foam depends largely upon the forces which exist during the expansion of the plastics; these are:

(a) the gas pressure which causes the material of the cell wall to flow as the volume of the cell increases;
(b) the viscoelastic reaction forces of the plastics which resist its flow.
(c) the surface tension forces which cause the flow of material from cell walls to the point at which they intersect.

The most favoured cell structure is one resulting in a minimum surface tension for the expanding plastics.

Density, when applied to a rigid plastics foam, refers to its bulk density, which is defined by the ratio: total weight/total volume of the plastics and gaseous components. Obviously, the gaseous component contributes considerably to the volume of the foam, whilst the solid plastics component contributes almost the entire weight.

The division of plastics, as given in section 3.2 is, of course, relevant to foams. As shown in Table 3.4 phenolic and polyurethane foams are thermosetting plastics which, apart from the polyester and epoxy resins already discussed, are the most commonly used in the construction industry. Expanded unplasticized PVC which is a thermoplastic material is also used in the industry.

6.3 Phenolic foam (P.F.)

Foamed phenol-formaldehyde resin (phenolic foam) is available in boards and slabs to BS 3927 [6.1]. It is a rigid thermosetting resin having a good chemical and temperature resistance, as well as a high resistance to water vapour transmission and water uptake; these properties are carried over into the foam.

The range of densities available at present varies from 30 kg/m³ up to 150 kg/m³. For particular applications, large cell structures can be provided or high density foams (up to 500 kg/m³) can be produced. Generally, phenolic foams are available in rectangular block form up to 2000 mm × 1000 mm × 500 mm and in square blocks up to 1000 mm × 1000 mm × 500 mm.

The foam achieves the highest classifications of the BS 476 fire tests Parts 5 and 7 and produces an optical smoke obscuration to Part 9 (see section 5.9.1.3) of less than 5%, compared with between 50 and 90% for most commercial grades of polystyrene and polyurethane foam. Phenolic foam burns in a similar way to natural timber, giving off only carbon dioxide, carbon monoxide and water, without the intermediate evolution of carbon laden smoke. As the combustion temperature progressively increases to above 200° C, the foams undergo gradual oxidation to become a carbonaceous structure, without any intermediate softening or dripping or any significant smoke, fume or flame production. The resulting rigid foam-char protects the remaining structure from the primary heat source.

Because of the inert nature of phenolic foam, only concentrated acids and alkalis can attack it chemically. It resists the solvents used in the adhesives and varnishes which are currently employed in building construction and it is claimed by one manufacturer that, despite its acid catalyst, it presents no particular corrosive problems with respect to metals. It is able to resist attack by fungi, bacteria, insect and rodent sources, even after prolonged exposure.

6.4 Rigid polyurethane foam

Chemically, rigid polyurethane foam is the most complex of all the plastics foams. The basic raw materials of polyurethane foam are a polyol, a polyisocyanate, a blowing agent, a catalyst and a surfactant. The resulting thermosetting foam can be made with either an open or closed cell structure.

When the inert organic liquids blend with the resin mixture an even and controlled foaming is produced. The pores of the foam are filled with a gas mixture which has a low value of conductivity and which is less than that of still air. The cell walls do not readily allow the gas to diffuse through them. In service there may be times in which the foam has free access to air and it may diffuse inwards until an equilibrium condition is reached, at which point the conductivity of the foam will be that of still air.

The main advantages of polyurethane over other foams lie in its excellent low thermal conductivity (0.02 W/m²/° C), good high temperature resistance (up to 120° C), low vapour permeability and the property of insitu foaming.

The behaviour of the foam in fire is not good although flame retardant grades are available by adding halogens at the time of mixing.

Coolag Ltd. is one of the main producers of polyurethane in the U.K. On small coupons of their standard material, they have given typical fire resistance properties, in the form of extent of burn, as 50 mm. This test is laid down in BS 4735, 1971 [6.2]. According to ASTM D1692–68 [6.3], the material is self extinguishing with an extent of burn of 50 mm.

Coolag Ltd. also produce an isocyanurate foam called Nilflam which is manufactured, in the same way as the standard material, on a continuous conveyor system, and allowed a free rise to provide a uniform cell structure. The above tests on small coupons of this material give nil and 5 mm extent of burn respectively.

Typical mechanical properties for the above two types of polyurethane foam are given in Table 6.1.

Table 6.1 Typical mechanical properties of standard and Nilflame polyurethane foams manufactured by Coolag Ltd.

		Standard foam	Nilflame isocyanurate foam
Density	(kg/m³)	29	32
Compressive strength			
in direction of rise	(kN/m²)	172	172
across rise	(kN/m²)	124	96
Tensile strength	(kN/m²)	276	206
Shear strength	(kN/m²)	172	138
Temperature limit	(°C)	110	150

The creep of rigid polyurethane foam, under static loading, has been studied by the Union Carbide Corporation of America and they have reported [6.4] that, according to their findings, the percentage of creep can be expressed in the form:

$$\text{creep}\,(\%) \;=\; XT/(AT + B)$$

where T = time

X = ratio of applied stress to elastic modulus of the foam

A and B = constants depending upon the temperature.

6.5 Urea formaldehyde (U.F.)

Although urea formaldehyde is not a structural material, it is relevant to introduce the foam here because of its applications in the construction industry. It is produced from a liquid system by mixing resin, water and a foaming hardener in a special machine. The foam leaves the mixing machine head or gun in a relatively soft fluid state, but quickly hardens into a rigid cellular mass. The foam has a density of 8 kg/m³, and is one of the lightest plastics cellular materials.

6.5.1 Uses of urea formaldehyde

In the building industry advantage is taken of the good heat insulating properties of the foam (the thermal conductivity value is 0.038 W/m²/°C, although this is not as low as that of polyurethane foam) by using it to infill the brickwork of the cavity walls. A criticism of this technique is that water may penetrate the inner walls, causing damp and the possibility of long term structural damage. Water transmission from the outer to the inner walls could result from the presence of fissures that form when the foam cures and which, if horizontal, act as a bridge. Tests undertaken by the foam suppliers and installers, in collaboration with the Building Research Establishment, have shown that whilst fissures do occur, they can be minimized by the correct controlled injection techniques. It was also shown that shrinkage-on-cure led to the creation of a small but nevertheless significant air space between wall and foam to maintain the water barrier.

6.6 Expanded PVC foam

Expanded PVC is a thermoplastics foam based on polyvinyl chloride resins or copolymers of polyvinyl chloride. It can be produced either by a mechanical blowing process or by one of several chemical blowing processes. It has an almost completely closed-cell structure and therefore a low water vapour transmission and low water absorption. This foam plastics tends to be expensive and therefore its use in the construction industry is limited to specific applications. It tends to be stronger and more rigid than the above mentioned cellular plastics and its low vapour transmission is an advantage when condensation might be a problem. Because of its rigidity, it is often used in sandwich constructions to increase the stiffness of the composite; however, care must be exercised if it is used as a building material because it tends to collapse in fire (although it burns with difficulty).

As an example of rigid expanded PVC products, B.T.R. Industries Ltd. make four standard grades and the low density material has the following advantages consistent with plastics materials:

(a) it has high tensile shear and compressive strength;
(b) it does not crumble under impact or vibration;
(c) it has good thermal insulation;
(d) it has low water vapour permeability;
(e) it is resistant to termites and bacterial growth;
(f) it has excellent chemical resistance.

Typical mechanical properties for the four standard grades are given in Table 6.2.

Table 6.2 Typical mechanical properties of standard grade PVC foams manufactured by B.T.R. Industries Ltd.

Properties		Standard grades			
Density	(kg/m³)	40	55	75	100
Compressive strength	(MN/m²)	0.34	0.48	0.93	1.38
Compressive modulus	(MN/m²)	10.34	17.23	19.30	27.58
Shear strength	(MN/m²)	0.34	0.48	0.93	1.14
Shear modulus	(MN/m²)	6.89	8.27	20.68	27.58
Flexural strength	(MN/m²)	0.38	0.55	1.03	1.38
Tensile strength	(MN/m²)	0.48	0.55	1.00	1.72

6.7 Mechanical testing of rigid plastics foams

Foamed plastics are approximately isotropic and their strength and stiffness are very approximately proportional to density. As a group, these plastics are unlikely to find application in high efficiency sandwich structures, although the stiffness of the continuum forming a folded plate is considerably increased with the utilization of a sandwich construction of chopped strand mat faces and foamed plastics core.

Tables 6.1 and 6.2 give typical values of the mechanical properties of some foamed plastics but it is usually necessary to undertake laboratory tests to determine the strength and stiffness of the particular form of core material to be used.

The properties which are generally required are:

(a) compressive strength;
(b) flat-wise tensile strength;
(c) shear tests.

These latter tests may be divided into three basic types [6.5]:

(a) in plane shear, in which the shear distortion takes place entirely in the plane of the composite material;
(b) twisting shear, in which the cross section of the composite (in bar or thin-sheet form) undergoes á twisting-type shear deformation;
(c) thickness shear, sometimes called transverse shear or interlaminar shear (in the case of laminates), in which the composite material sheet undergoes shearing deformation in a plane normal to the plane of the sheet.

It is the first shear type that will be described in this book and which is of importance when considering sandwich beams and slab construction.

6.7.1 Compressive strengths

It is necessary to determine the compressive characteristics of rigid plastics foams used as core materials in structural constructions. The characteristics are determined in directions perpendicular and parallel to the plane of the facings corresponding to directions T and L of Figure 6.1; this is how the foam core would be placed in an actual sandwich. The test procedure is laid down in ASTM designation C365 [6.6].

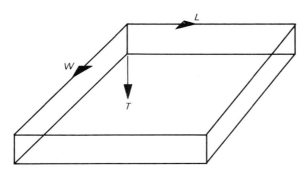

Figure 6.1 Planes within a core material of sandwich construction

The significance of the test is that it produces information on the behaviour of rigid plastics foams under uniaxial compressive loads:

$$\text{compressive stress } \sigma_c = \frac{\text{compressive load}}{\text{cross sectional area}}$$

6.7.2 Flatwise tension test

This method covers the procedure for determining the tensile properties of rigid plastics foams for use as core material in sandwich construction. Properties are determined in a direction normal to the plane of the facings as the foam core would be placed in the sandwich construction in direction T and corresponding to the rise direction of the foam. Basically, the test requires the specimen to be bonded between two steel loading plates which are then secured onto two loading blocks. The loading blocks and plates must be sufficiently stiff to keep the bonded surface as flat as possible.

American Standard Test Methods for determining the tensile properties of rigid plastics foams are given in references [6.7] and [6.8]. These standards require specimens which are lathe turned to specific dimensions. The significance of the test lies in the information it provides on the behaviour of rigid plastics foams when loaded in tension. In the design of sandwich construction, a knowledge of the ultimate tensile strength of the core material is usually sufficient; it will help to decide whether the face wrinkling will resist either the facing tearing from the core or buckling into the core.

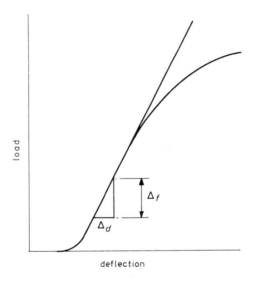

Figure 6.2 Typical flatwise tension test result

A typical recording of load versus deflection is shown in Figure 6.2.

$$\text{tensile stress } \sigma \ = \ \frac{\text{tensile load}}{\text{cross sectional area}}$$

6.7.3 Shear test

This method covers the procedure for determining the shear properties of rigid plastics foams for use as core materials in structural sandwich constructions. Figure 6.3 gives details of single shear tests for determining the shear strength and stiffness in the LT and TW planes (Figure 6.1). Samples of the core material are bonded to two thick plates and loaded in the direction as

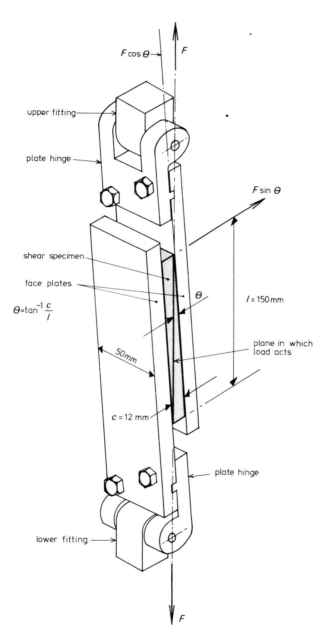

Figure 6.3 Shear test for determining shear strength of foam core

shown in Figure 6.3. The standard shear test method is laid down in A.S.T.M. designation C 273–61 [6.9] and a typical recording of load against deflection for a shear test is given in Figure 6.4.

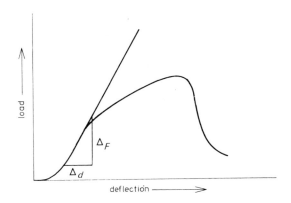

Figure 6.4 Typical curve of load versus deflection for shear test

If the shear stress across the width of the specimen is assumed constant, then:

$$\text{shear stress } \tau = \frac{\text{shear load}}{\text{area of shear}}$$

$$= \frac{F \cos \theta}{L\, b}$$

$$\text{shear strain } \gamma = \frac{r}{c}$$

where r = movement of one face relative to the other
c = thickness of specimen

$$\text{shear modulus} = G = \frac{c\, F \cos \theta}{L\, b\, r}$$

6.8 Sandwich beams

The sandwich beam is defined in Appendix A and is manufactured by sandwiching a core of low density material of thickness c between two faces of thickness t. It is likely that the beam will be structurally efficient only if the faces are much stronger and stiffer than the core material and if the density of the core material is much lower than that of the faces; these conditions are usually met in practice.

The flexural rigidity EI of a sandwich beam, which is usually given the symbol D, is the sum of the flexural rigidities of the two separate parts of the composite, namely the core and the faces, measured about the centroidal axis of the whole section. From Figure 6.5 it will be clear that

$$D = E_f \frac{bt^3}{6} + E_f \frac{btd^2}{2} + E_c \frac{bc^3}{12} \tag{6.1}$$

where E_f and E_c are the modulus of elasticity of the faces and core respectively.

The first term on the right hand side of the equation represents the local stiffness D_f of the faces bending about their own centroidal axes. The third term represents the bending stiffness of the core. In practice the second term is the dominant one and, provided the first and third terms are less than 1% of the second, then they must be discounted.

Therefore, provided:

$$[(E_f\, bt^3)/6]\,/\,[(E_f\, btd^2)/2] < \frac{1}{100}$$

and $[(E_c\, bc^3)/12]\,/\,[(E_f\, btd^2)/2] < \dfrac{1}{100}$

i.e. $\dfrac{d}{t} > 5.77$ and $\dfrac{E_f}{E_c}\dfrac{t}{c}\left(\dfrac{d}{c}\right)^2 > \dfrac{100}{6}$ \hfill (6.2)

$$D = E_f \frac{btd^2}{2} \tag{6.3}$$

Under these conditions, the stresses in the faces may be determined by the use of ordinary bending theory. As the sections remain plane and perpendicular to the longitudinal axis, the strains at any point distance z below the centroidal axis are Mz/D. The sign convention for bending of beams is given in Figure 6.5

The stresses in the faces $= \dfrac{Mz}{D}\, E_f$

$(\sigma_f)_{max} = \pm \dfrac{Mh\, E_f}{2D}$ \hfill (6.4)

Similarly the stresses in the core $= \dfrac{Mz}{D}\, E_c$

$(\sigma_c)_{max} = \pm \dfrac{Mc\, E_c}{2D}$ \hfill (6.5)

6.8.1 Shear stress in core

The assumptions in the ordinary bending theory lead to the shear stress distribution in a homogeneous beam at a depth z below the neutral axis as:

$$\tau = \frac{Q\,a\,\bar{z}}{b\,I} \qquad\qquad (6.6)$$

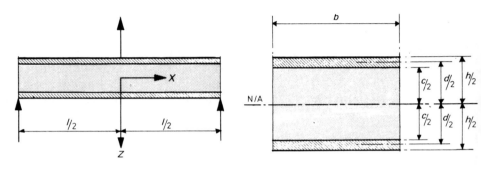

(a)

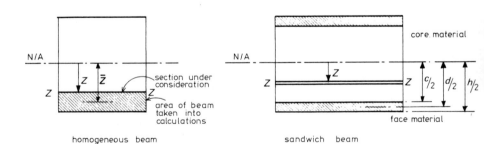

section under consideration

area of beam taken into calculations

core material

face material

homogeneous beam

sandwich beam

(b)

M positive

Q positive

(c)

where Q = shear force

a = area of the section of the beam at section under consideration (see Figure 6.5)

$\bar{z}$ = distance of centroid of area a to centroidal axis of beam

b = width of beam at section under consideration

I = second moment of area of whole section of beam.

For a composite sandwich beam the shear stress in the core is:

$$\tau = \frac{Q}{Db} \Sigma a\, \bar{z}\, E \qquad (6.7)$$

where D is the flexural rigidity of the entire section

E is the modulus of elasticity of the various component sections above section under consideration.

Considering Figure 6.5

$$\Sigma a\, \bar{z}\, E = \left[E_f\, b\, t\frac{d}{2}\right] + \left[\frac{c}{2} - z\right]\left[\frac{c}{2} + z\right]\frac{b}{2} E_c$$

$$= E_f \frac{b\, t\, d}{2} + \left[\frac{c^2}{4} - z^2\right]\frac{b\, E_c}{2}$$

therefore $\tau = \dfrac{Q}{D}\left[E_f \dfrac{t\, d}{2} + \dfrac{E_c}{2}\left\{\dfrac{c^2}{4} - z^2\right\}\right] \qquad (6.8)$

The maximum core shear stress will occur at $z = 0$, and the minimum core shear stress will occur at $z = \pm\, c/2$.

Figure 6.5 Sandwich beam dimensions and sign convention; (a) dimensions of sandwich beam for flexural rigidity equations; (b) dimensions for shear stress equations; (c) the sign convention adopted for the stress resultants is shown. The positive direction of the external forces required to balance the stress resultants at the cut section is derived from a right-handed system of coordinates x, y and z. Consequently the relationship between the curvature and bending moment is:

$$M = -EI\frac{d^2z}{dx^2} \qquad \text{deflection } = w \qquad \text{moment } M = -D\frac{d^2w}{dx^2}$$

$$Q = \frac{dM}{dx} \qquad \text{slope } = \frac{dw}{dx} \qquad \text{shear force } Q = -D\frac{d^3w}{dx^3}$$

$$q = -\frac{dQ}{dx} \qquad \text{curvature } = \frac{d^2w}{dx^2} \qquad \text{distributed load } q = D\frac{d^4w}{dx^4}$$

H

If the ratio of the maximum core shear stress to the minimum core shear stress is within 1% of unity,

$$\text{i.e.} \quad \left[1 + \frac{E_c}{E_f} \frac{c}{t} \frac{c}{d} \frac{1}{4} \right] \quad \text{where} \quad 4 \frac{E_f}{E_c} \frac{t}{c} \frac{d}{c} > 100$$

it may be assumed that the core is too weak to contribute to the shear stress and this stress may be considered constant across the core; the last term in the right hand side of equation (6.8) is then ignored and the constant shear stress in the core is

$$\tau = \frac{Q}{D} \frac{E_f t d}{2}$$

If in addition to the above, the condition associated with equation (6.2) is also satisfied, then the shear stress in the core is

$$\tau = \frac{Q}{b d} \tag{6.9}$$

Figure 6.6 shows the shear stress distribution in the sandwich beam.

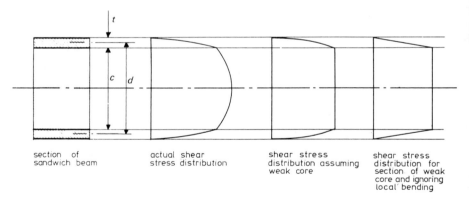

section of   actual shear   shear stress   shear stress
sandwich beam   stress distribution   distribution assuming distribution for
                                          weak core   section of weak
                                                                core and ignoring
                                                                local' bending

Figure 6.6 Shear stress distribution for sandwich beam

In using the equations (6.4), (6.5) and (6.9) for the stresses in bending and shear, the beam is defined as a thin face one (see definition in Appendix, p. 219) and the following assumptions have been made:

(1) A core with a low modulus resulting from a cell structure is considered to be a continuous elastic material.

(2) In sandwich constructions of G.R.P. faces and foam cores, the faces are much stronger, stiffer and denser than the core. Therefore, the contribution of the core to the overall stiffness of the sandwich construction is small

and may be neglected. Consequently, the direct stresses which arise in the core under bending are small and may also be neglected; the shear stresses in planes perpendicular to the faces are uniform across the thickness of the core.

(3)   The core is assumed to be sufficiently stiff in the direction normal to the faces to maintain them the correct distance apart; this assumes stiffening members are placed under concentrated loads.

6.8.2 Thin face sandwich beams

The total deflection w of the beam can be regarded as the sum of the primary deflections (bending deflection) w_1 and the secondary deflections (shear deflection) w_2. The primary deflection may be calculated by the ordinary theory of bending and the secondary deflection also by this theory based on the combined flexural rigidity of the faces as they bend locally about their own separate centroidal axis.

The central deflection of a sandwich beam under a central point load can be represented as shown in Figure 6.7a. The primary displacement occurs when plane cross sections remain plane and perpendicular to the longitudinal axis of the beam. This is illustrated in Figure 6.8 by a short length of beam where $abcdefg$, which was straight in the undeformed sandwich, remains straight and perpendicular to the axis of the beam after bending.

The rotation of this line dw_1/dx gives the slope of the beam and the stresses are related to the displacements by the simple theory of bending. Similarly, Figure 6.8 also shows a short length of beam which has undergone a secondary displacement which occurred when the faces bent about their own individual centroidal axes but underwent no axial extension or contraction. The lines, abc and efg, are perpendicular to the longitudinal axis of the beam. The rotation of these lines is equal to the slope of the beam dw_2/dx. Therefore,

$$\frac{dw_2}{dx} = \gamma \frac{c}{d}$$

$$= \frac{Q}{b\,d\,G_c}\frac{c}{d} \qquad\qquad (6.10)$$

$$= \frac{Q}{A\,G_c} \quad \text{where } A = \frac{bd^2}{c}$$

and G_c is the modulus of rigidity of the core.

The deflection due to shear at the centre of the beam carrying a single point load at this position is obtained by integrating equation (6.10) and is:

$$w_2 = \frac{W}{4\,A\,G_c} \qquad\qquad (6.11)$$

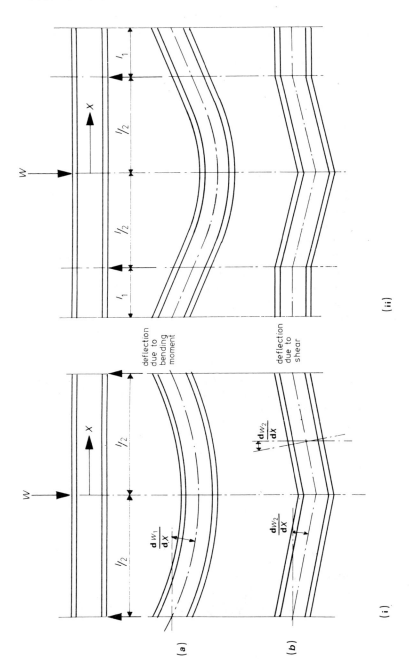

Figure 6.7 Deflection of sandwich beam under centrally applied load: (i) sandwich beam without overhang but with weak core; (ii) sandwich beam with overhang and weak core

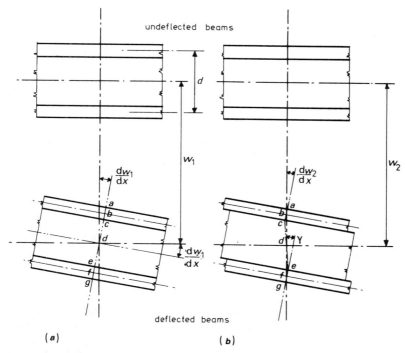

Figure 6.8 Primary and secondary deflections of short length beam; (a) primary displacement (see also Figure 6.7i and ii (a)); (b) secondary displacement (see also Figure 6.7i and ii (b))

so the total deflection due to bending and shear for this beam is:

$$w = \frac{WL^3}{48\,D} + \frac{WL}{4\,G_c\,A} \tag{6.12}$$

Generally, for a thin face beam under a statically determined and symmetrically loaded condition, the total deflection may be determined by the addition of the primary and secondary deflections calculated separately, utilizing in each case the total load on the beam. To calculate the deflections for very thin faces or thick faces it is necessary to use modifications of the above formula.

6.8.3 Thick face sandwich beams

For thick face sandwich beams the first term on the right hand side of equation (6.1) cannot be ignored. The flexural rigidity of the beam is then:

$$D = E_f \frac{b\,t\,d^2}{2} + E_f \frac{b\,t^3}{6} \tag{6.13}$$

The shear stiffness D_Q has been defined[6.10] as the shear force which must be applied to the beam to produce a unit slope in the secondary mode of deformation. If dw_2/dx is equal to unity, then $\gamma = d/c$ and the shear stress in the core is $G_c\, d/c$ and the shear force carried by the whole cross section of the beam is:

$$D_Q = \left[G_c\, \frac{d}{c}\right] b \left[\frac{t}{2} + c + \frac{t}{2}\right] = G_c\, \frac{bd^2}{c} \tag{6.14}$$

6.8.4 Characteristic behaviour of a sandwich beam

A sandwich beam under load exhibits characteristic behaviour which depends upon the three non-dimensional parameters: D_f/D (the ratio of the face stiffness to the total stiffness of the sandwich), $D/L^2 D_Q$ (the ratio of the total flexural stiffness to the product of the shear stiffness and the square of the span), and L_1/L (the ratio of the length of overhang and length of beam).

The total deflection due to bending and shear at the centre of the beam carrying a single point load at this position and having an overhang L_1 (see Figure 6.7b) is:

$$w = \frac{WL^3}{48\,D} + \frac{WL}{4\,D_Q}\left[1 - \frac{D_f}{D}\right]^2 S_1 \tag{6.15}$$

(The value of S_1 is given in Appendix 6.1)
This is the general formula from which equation (6.11) may be derived.

If w is normalized with respect to w_1, where w and w_1 are defined above, the dimensionless quantity r_1, is obtained:

$$r_1 = 1 + \frac{12D}{L^2\,D_Q}\left[1 - \frac{D_f}{D}\right]^2 S_1$$

Allen [6.10] has drawn a complete family of curves of r_1 versus $D/L^2 D_Q$, each curve representing a particular value of D_f/D; these curves define only the deflection of a sandwich beam with a central point load and no overhang. These curves are shown in Figure 6.9.

Allen shows that:

(a) At one limit of the relationship, where $D/L^2 D_Q < 0.01$, r_1 approaches unity, the span L is large and $w = WL^3/48D$ and the deformation is almost wholly due to the primary mode.

(b) At the other limit of the relationship where $D/L^2 D_Q > 1000$, r_1 approaches the value D/D_f, the span L is short and

$$w = \frac{WL^3}{48\,D_f} \tag{6.16}$$

(c) Between these two limits ($0.01 < D/L^2 D_Q < 1000$) the deformation is a combination of the primary and secondary deflections and the stresses in

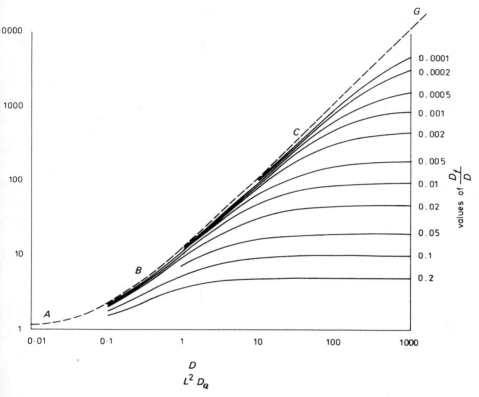

Figure 6.9 Family of curves for r_1 plotted against parameter $D/L^2 D_Q$ (based on Figure 6.5 of ref. [6.10])

the beam are complicated. In Figure 6.9, the curve ABCG has been plotted for the result of a thin face sandwich and it is evident that for $D/L^2 D_Q < 10$ the curve essentially defines the sum of the deformation due to ordinary bending and the deformation due to shear strain in the core.

Similar graphs to that shown in Figure 6.9 can be constructed to illustrate the way in which deflections, bending moments, shear forces and critical loads depend upon D_f/D and upon $D/L^2 D_Q$ for a wide range of sandwich beams. Allen [6.11] has given an outline of the method of solution with formulae, to calculate the coefficient for beams with central point load and uniformly distributed load. The formulae and coefficients for a simply supported beam with a central point load and with a uniformly distributed load have been reproduced at the end of this Chapter (Appendix 6.1).

In the analysis of sandwich constructions involving narrow beams (a narrow beam is defined as one in which b is less than the core depth c.

Conversely, a wide beam is one in which b is much greater than c) the lateral expansions and contractions of faces take place without causing unduly large shear strains in the core. The stress condition is therefore in a state of unidirectional stress and the ratio of stress to strain is equal to E. When local bending stresses are set up in the face of the sandwich, they act as thin plates in cylindrical bending and the ratio of stress to strain in this case is $E/(1-v^2)$. However, as these strains are usually of secondary importance, the ratio may be taken as E without undue inaccuracy. For wide beams the strains in the lateral directions are restrained by the core and therefore are assumed to be zero. In this case the ratio of the longitudinal stress to strain is $E/(1-v^2)$.

6.9 Buckling of sandwich struts with thin faces

The maximum axial compressive force which a pin-ended elastic strut can support before it becomes unstable is equal to the Euler load P_E given by:

$$P_E = \frac{\pi^2 D}{L^2} \qquad (6.17)$$

where D = flexural rigidity

In a sandwich strut in which the core material is of low modulus, it is likely that shear deformations within the core will occur; these reduce the stiffness of the strut member.

For a thin face and an antiplane core (see definition in Appendix, p. 219), the flexural rigidity is given by:

$$D_1 = E_f b t d^2/2$$

When the strut deflects, the shear stress distribution is similar to that given in Figure 6.6 and the total displacement is the addition of the displacement w_1 due to ordinary bending and that due to shear deformation w_2 of the core.

The buckled strut is shown in Figure 6.10 and at any cross section the moment is equal to

$$P(w_1 + w_2) = -D_1 w_1''$$

Reference [6.11] has shown that this equation yields a differential equation in w_1.

$$w_1''' + \alpha^2 w_1' = 0$$

The solution of this equation with correct boundary conditions (i.e. $w_1 + w_2 = 0$ at $x = 0$ and at $x = L$) leads to the critical load of the sandwich strut.

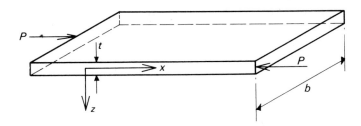

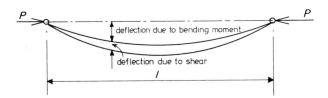

Figure 6.10 Deformations of buckled strut

$$P = \frac{P_{\mathrm{E}}}{1 + P_{\mathrm{E}}/AG_{\mathrm{c}}} \; ; \; P_{\mathrm{E}} = \pi^2 \, D_1/L^2 \qquad (6.18)$$

In the above formulae for sandwich struts with thin faces it has been assumed that the strut is narrow and that it bends anticlastically. If the strut is wide and bends cylindrically the value for E_{f} should be replaced by $E_{\mathrm{f}}/(1-\nu_{\mathrm{f}}^2)$ (see definition in Appendix, p. 219).

6.10 Buckling of sandwich struts with thick faces

In the case of a pin-ended sandwich strut with thick faces, Allen [6.11] has shown that the critical load is:

$$P_{\mathrm{cr}} = P_{\mathrm{E}} \left[\frac{1 + (P_{\mathrm{Ef}}/P_{\mathrm{c}}) - (P_{\mathrm{Ef}}/P_{\mathrm{c}}) \, (P_{\mathrm{Ef}}/P_{\mathrm{E}})}{1 + (P_{\mathrm{E}}/P_{\mathrm{c}}) - (P_{\mathrm{Ef}}/P_{\mathrm{c}})} \right] \qquad (6.19)$$

where P_{E} $= \pi^2 D/L^2$ = Euler load of strut when there are no shear strains in the core

P_{Ef} $= \pi^2 D_{\mathrm{f}}/L^2$ = sum of the Euler loads of the faces, considered as two independent struts

P_{c} = critical load for shearing instability.

Equation (6.19) may be rewritten as:

$$\frac{P_E}{P_{cr}} = \frac{1 + \pi^2 \dfrac{D}{L^2 D_Q} \left[1 - \dfrac{D_f}{D}\right]}{1 + \pi^2 \dfrac{D}{L^2 D_Q} \left[1 - \dfrac{D_f}{D}\right] \dfrac{D_f}{D}} \tag{6.20}$$

P_E/P_{cr} is the ratio of the Euler load of a strut (of the same dimensions as the sandwich strut) to the critical load of the sandwich strut, and if this ratio is plotted against $D/L^2 D_Q$, for various values of D_f/D, it may be shown that the results resemble those given in Figure 6.9 where P_E/P_{cr} is substituted for r_1. This means that the true ratios $D/L^2 D_Q$ and D_f/D can be calculated for any situation, and by referring to Figure 6.9 it will be evident which type of sandwich behaviour can be expected. The same reasoning may be applied when considering the buckling of sandwich beams as was applied to the bending of sandwich beams in section 6.8.4.

(a) If $D/L^2 D_Q < 0.01$, ordinary bending theory may be used and shear deflection ignored.
(b) If the calculated values of $D/L^2 D_Q$ and D_f/D are such that the plotted points lie on or near the line ABCG, local bending of the faces is not likely to be important. The beam or column will then have ordinary bending deflections in addition to deflection due to shear in the core.
(c) If the plotted points lie well below the line ABCG, then local bending of the faces is expected and the thick face theory must be used.

6.11 Wrinkling instability of faces of sandwich struts with cores of finite thickness

In addition to the critical load on a sandwich strut there is an additional component which could cause failure of the strut by introducing a local instability, known as *wrinkling,* and which is associated with the short wavelength ripples in the faces. Wrinkling may also be present in the compression faces of sandwich beams. The short wavelength wrinkling instability of the faces could occur at a lower axial load than the critical buckling load. Analyses of this problem have been made by Gough, Elam and De Bruyne [6.12] and by Hoff and Mautner [6.13].

The wrinkling theory refers only to local bending of the faces and ignores completely their membrane strains. On the other hand, it will be realized that ordinary buckling theory is concerned only with the overall stability of the sandwich component. Wrinkling and other forms of local instability of sandwich beams and struts have been well documented by Allen [6.11]. The theory is too extensive to quote here and therefore only the results for use by designers will be presented.

There are three types of wrinkling instability:

(a) the sandwich beam develops wrinkling in the compressive face; the tensile face remains perfectly flat;
(b) the antisymmetrical wrinkling in sandwich struts;
(c) the symmetrical wrinkling in sandwich struts.

These three types of wrinkling have been shown in Figure 6.11.

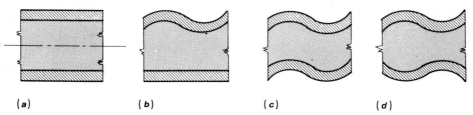

(a)   (b)   (c)   (d)

Figure 6.11 Wrinkling instability; (a) undeformed faces; (b) wrinkling occurring in compressive faces; tensile face undeformed; (c) antisymmetric wrinkling of sandwich beam; (d) symmetric wrinkling of sandwich beam; (based on Figure 8.4 of ref. [6.11])

The stress at which wrinkling commences in the faces of axially loaded struts and panels with isotropic cores such as phenolic or polyurethane foamed plastics (which for design purposes are assumed isotropic), is given by:

$$\sigma = B E_f^{1/3} E_c^{2/3} \tag{6.21}$$

where E_f = modulus of elasticity of the strut face material

   = $E_f / (1 - \nu_f^2)$ for wide panels

   E_c = modulus of elasticity of the core material

   B = buckling coefficient.

Hoff and Mautner [6.13] have quoted an empirical value for B of 0.4. This must only be considered an approximate quantity.

6.12 Buckling of sandwich panels

This section gives the essential equations for determining the critical edge force per unit length of sandwich panels. The panel considered will be of the form shown in Figure 6.12 the notation is as for sandwich beams. It is supported on four sides and bends in two directions.

The assumptions made in the following discussions are:

(a) deflections are small; therefore the ordinary theory of bending is valid;
(b) there is no extension of the middle section of the panel, therefore the membrane forces are unaffected by displacements in the z-direction;
(c) as the faces are thin, local bending stiffness within them is neglected.

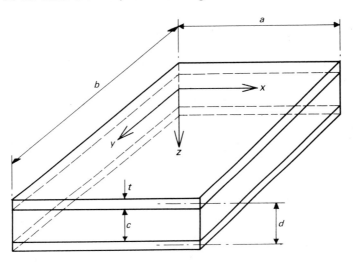

Figure 6.12 Sandwich panel with equal faces $a =$ length of panel $b =$ width of panel

6.12.1 Buckling of thin face sandwich panels

The critical buckling edge load per unit length in the x-direction for a thin face sandwich is:

$$P_{cr} = \frac{\pi^2 D_2}{b^2} K \qquad \text{(this load causes buckling in the} \qquad (6.23)$$
$$m, n \text{ mode)}$$

where $D_2 = \dfrac{E_f t d^2}{2(1-v_f^2)}$ = flexural rigidity of the panel per unit length (see definition in Appendix, p. 219)

b = width of panel

K = dimensionless coefficient

$$= \left\{ mb/a + (n^2 a/mb) \right\}^2 \bigg/ \left\{ 1 + \rho \left[(m^2 b^2/a^2) + n^2 \right] \right\}$$

a = length of panel

$$\rho = \frac{\pi^2}{b^2} \left[\frac{E_f d^2 \ t}{2(1-v_f^2)} \right] \left[\frac{c}{G_c \ d^2} \right]$$

$$= \frac{\pi^2}{2(1-v_f^2)} \frac{E_f}{G_c} \frac{t\,c}{b^2}$$

$G_c d^2/c$ = shear rigidity of the panel per unit length

or $\rho = \dfrac{\pi^2}{b^2}$ times ratio of flexural rigidity to the shear rigidity.

The dimensionless coefficient K may be obtained from Figure 6.13, knowing the ratio a/b and ρ. Equation (6.23) may be used for an isotropic face (*viz.* a C.S.M. laminate) and an assumed isotropic foamed plastics core.

To determine the buckling load of a panel with thin or very thin faces, it is necessary to find the mode which gives the smallest critical load. For any value of m this critical load value will exist when $n=1$. Therefore, the dimensionless coefficient K may be obtained from Figure 6.13 when values of a/b and ρ are known. It should be noted that the curves in this figure ignore the effect of D_f on the secondary deformation and they can only be relied upon if D_f/D and $D/L^2 D_Q$, for the plate, combine to give a point in Figure 6.9 which is close to the line ABCG.

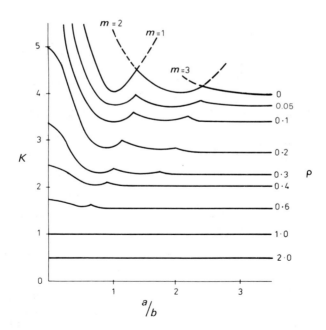

Figure 6.13 Relationship between K and a/b for known values of p (equation (6.23)) (based on Figure 5.4 of ref. [6.11])

6.12.2 Buckling of thick face sandwich panels

The critical buckling edge load per unit length in the x-direction for a thick face sandwich is:

$$P_{cr} = \frac{\pi^2 D_2}{b^2} K_t \qquad (6.24)$$

where P, a, b, D_2 are as defined in equation (6.23).

$$K_t = K + \left[\frac{mb}{a} + \frac{n^2 a}{mb}\right]^2 \left[\frac{t^2}{3d^2}\right]$$

K = dimensionless coefficient in equation (6.23).

The dimensionless coefficient K_t may be obtained from Figure 6.14 when the values of a/b and ρ are known and when $t/d = 0$ and 0.3. These values of t/d represent the extreme values of any practical sandwich.

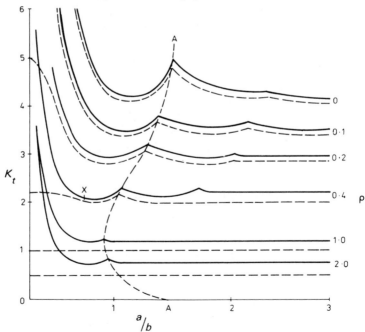

Figure 6.14 Buckling coefficient K_t in equation (6.24). Simply-supported isotropic sandwich panel with thick faces and uniform edge load in the x-direction. Broken and full lines represent the extreme cases $t/d = 0$, 0.3, respectively. (Based on Figure 6.3 of ref. [6.11], by permission of Pergamon Press)

Equation (6.24) may be utilized for an isotropic face and isotropic core (i.e. faces of a chopped strand laminate and an assumed isotropic foamed plastics core).

6.13 Bending of simply supported panels with uniform transverse load

The maximum transverse deflection of a sandwich panel, consisting of an isotropic thin face and isotropic core, due to a uniform transverse load q per unit area is given by reference [6.11] as:

$$w_{max} = \frac{q\,b^4}{D_2}\left[\beta_1 + \rho\beta_2\right] \tag{6.25}$$

where b = width of the panel in the y-direction

$\qquad D_2 = E_f\,t\,d^2\,/\,2(1-\nu_f^2)$

$\qquad\qquad \rho$ = flexural and shear rigidity ratio as given in equation (6.23)

$\quad \beta_1$ and β_2 = coefficients which are given in Figure 6.15.

Timoshenko and Woinowsky-Krieger [6.14] have given an equivalent formula, equation (6.26), for a homogeneous plate. It is necessary to obtain the numerical factor α which depends upon the ratio a/b of the sides of the plate; the value of the maximum deflection computed from equation (6.26) yields a similar result to that obtained from equation (6.25).

$$w_{max} = \frac{q\,b^4}{D}\,\alpha \tag{6.26}$$

where α may be obtained from Chapter 5, Table 8 of ref. [6.14].

The formulae quoted above should be used only when small transverse deflections of the panel take place. If large deflections are experienced then ref. [6.15] should be referred to.

Corresponding membrane stresses in the faces and shear stresses in the core may be obtained from the following equations [6.11].

The direct stresses in the faces are:

$$\sigma_{xx} = \frac{q\,b^2}{dt}\left[\beta_3 + \nu\beta_4\right] \tag{6.27}$$

$$\sigma_{yy} = \frac{q\,b^2}{dt}\left[\beta_4 + \nu\beta_5\right] \tag{6.28}$$

The maximum stresses are located at the centre of the plate ($x = a/2, y = b/2$, Figure 6.12)

The shear stresses in the faces are:

$$\sigma_{xy} = \frac{q\,b^2}{dt} \left[1-\nu\right] \beta_5 \tag{6.29}$$

The maximum value is located at the corners of the slab ($x = 0, y = 0$).

The shear stresses in the core are:

$$\sigma_{zx} = \frac{q\,b}{d} \beta_6 \qquad \text{and located in the middle of the side of length } b \tag{6.30}$$
$$(x = 0, y = b/2)$$

$$\sigma_{yz} = \frac{q\,b}{d} \beta_7 \qquad \text{and located in the middle of the side of length } a \tag{6.31}$$
$$(x = a/2, y = 0)$$

where d and t have been defined in Figure 6.12 and β_3 to β_7 may be obtained from Figure 6.15.

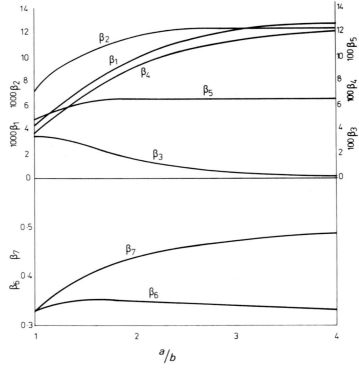

Figure 6.15 Values of $\beta_1 - \beta_7$ in equations (6.25) to (6.31). Isotropic panel with very thin faces (based on Figure 5.5 of ref. [6.11], by permission of Pergamon Press)

In the above equations the stresses are independent of the shear stiffness of the core and are identical with those obtained when the shear deformations in the core are neglected. Consequently the results given by Timoshenko and Woinowsky-Krieger [6.14] could also be used to determine the stresses in a sandwich plate.

6.14 Simply supported sandwich panels with edge load and uniform transverse load

Allen [6.11] has discussed this topic fully for general sandwich constructions, and Benjamin [6.16] has also quoted formulae which may be used to investigate the problem. However, the formulae derived by Allen incorporate a number of coefficients, all of which are dependent upon dimensionless parameters. It is recommended that reference be made to the above for information concerning edge load and uniform transverse load on simply supported sandwich panels.

6.15 Summary: a guide for design of sandwich beams and of struts and panels with edge loads

Figure 6.9 gives an important graphical result as it shows immediately which type of sandwich behaviour can be expected. It may be used as an approximate design guide when the parameter values $D/L^2 D_Q$ and D_f/D are known and where the quantity r_1 may be taken as an approximate estimate for:

(a) the ratio of (deflection of a sandwich beam of span L)/(deflection of simple beam of the same dimensions and load);
(b) the ratio of (Euler load of a simple strut of length L)/(the critical load of a sandwich strut of the same dimensions);
(c) the ratio of (deflection of a sandwich panel)/(deflection of an equivalent simple panel). The panel is supported on four sides, with the shorter span of length L and with a transverse load applied to it;
(d) the ratio of (critical edge load on a simple panel)/(critical edge load of an equivalent sandwich panel). The panel is supported on four sides with the shorter span of length L.

The graph may be interpreted as follows:

(a) When the ratio $D/L^2 D_Q$ is less than 0.01 (i.e. for very large spans or cores with high shear moduli) r_1 approaches unity. This implies that the shearing deformations of the core are not important and the panel may be analyzed by the application of the ordinary bending theory.
(b) If the calculated values of $D/L^2 D_Q$ and D_f/D give plotted points close to the curve ABCG, the behaviour of the member is as for an ordinary beam, column or panel with the addition of shear deflection in the core.

I

Most practical sandwiches with flat faces have values of $D_f/D \not> 0.01$ and $c/t \not< 5$ which will give points close to the curve ABCG provided $D/L^2 D_Q < 1$.

(c) If the plotted points lie close to the horizontal region of the D_f/D curve, then the composite action of the sandwich has been largely lost and the faces act as independent beams or columns.

Having established the dimensions of the sandwich component, it is advisable to undertake a laboratory test to verify the deflections.

Appendix 6.1

The total deflection due to bending and shear at the centre of a beam carrying a single point load at this position is:

$$w = \frac{WL^3}{48\,D} + \frac{WL}{4\,D_Q}\left[1 - \frac{D_f}{D}\right]S_1$$

The total deflection due to bending and shear at the centre of a beam carrying a uniformly distributed load is:

$$w = \frac{5\,q\,L^4}{384\,D} + \frac{q\,L^2}{8\,D_Q}\left[1 - \frac{D_f}{D}\right]^2 S_2$$

The coefficients S_1 and S_2 are defined as:

$$S_1 = 1 - \frac{\sinh\theta + \beta_1\,(1 - \cosh\theta)}{\theta}$$

$$S_2 = 1 + \frac{2\,\beta_2}{\theta}\,(1 - \cosh\theta)$$

$$\beta_1 = \frac{\sinh\theta - (1 - \cosh\theta)\tanh\phi}{\sinh\theta\,\tanh\phi + \cosh\theta}$$

$$\beta_2 = \frac{1}{\theta}\left[\frac{1 + \theta\,\tanh\phi}{\sinh\theta\,\tanh\phi + \cosh\theta}\right]$$

$$\theta = \frac{1}{2}\left\{\left[\frac{D}{L^2 D_Q}\right]\frac{D_f}{D}\left[1 - \frac{D_f}{D}\right]\right\}^{-1/2}$$

$$\phi = 2\theta\frac{L_1}{L}$$

The above equations are given for simply supported beams with overhangs of value L_1. The various symbols have been defined in the Chapter.

References

6.1. 'Phenolic foam materials for thermal insulations and building applications', BS 3927, British Standards Inst., London, 1965.

6.2. 'Laboratory methods of test for assessment of the horizontal burning characteristics of specimens no larger than 150 mm × 50 mm × 13 mm (nominal) of cellular plastics and cellular rubber materials when subjected to a small flame', BS 4735, British Standards Inst., London, 1974.

6.3. 'Test for flammability of plastic sheeting and cellular plastics', American Society for Testing Materials, Standard-designation D1692–68.

6.4. U.S. Department of State, 'The potential use of foam plastics for housing in underdeveloped areas', Architectural Research Laboratory, University of Michigan, Ann Arbor, 1963.

6.5. CHAMIS, C.C. (Ed.), *Structural Design and Analysis, Part II,* Vol. 8 of *Composite Materials,* BROUTMAN, L.J. & KROCK, R.H. (Eds.), Academic Press, 1975.

6.6. 'Flatwise compressive strength of sandwich cores', American Society for Testing Materials, Standard-designation C365.

6.7. 'Tension test of flat sandwich constructions in flatwise planes', American Society for Testing Materials, Standard-designation C297–61.

6.8. 'Tests for tensile properties of rigid cellular plastics', American Society for Testing Materials, Standard-designation D1623–64.

6.9. 'Shear test in flatwise plane of flat sandwich construction or sandwich core', American Society for Testing Materials, Standard-designation C273.

6.10. ALLEN, H.G., 'Analysis and design methods used in plastics sandwich panel construction' Chapter 6 in *Use of Load Bearing and Infill Panels,* (ed.) HOLLAWAY, L., Manning Rapley Publishing, 1976.

6.11. ALLEN, H.G., *Analysis and Design of Structural Sandwich Panels,* Pergamon Press, 1969.

6.12. GOUGH, G.S., ELAM, C.F. & DE BRUYNE, N.D., 'The stabilisation of a thin sheet by a continuous supporting medium', *J. Roy. Aero. Soc.,* **44,** 349, 1970, pp. 12–43.

6.13. HOFF, N.J. & MAUTNER, S.E., 'Buckling of sandwich-type panels, *J. Aero-Sci.,* **12,** 3, July, 1945, pp. 285–297.

6.14. TIMOSHENKO, S., and WOINOWSKY-KRIEGER, S., *Theory of Plates and Shells,* McGraw-Hill, 1959.

6.15. ALWAN, A.M., 'Large deflection of sandwich plates with orthotropic cores, *J. AIAA,* **2,** 10, October, 1964, pp. 1820–1822.

6.16. BENJAMIN, B.S., *Structural Design with Plastics,* Van Nostrand, 1969.

7 Structural forms for G.R.P. materials and their use in the construction industry

7.1 Introduction

It has been shown previously (Table 4.1) that plastics and glass fibre materials, and the composites manufactured from them, have low moduli of elasticity. Consequently, if such composites are to be used as load bearing components, the structural form must be chosen so as to overcome the apparent lack of stiffness in the overall structure. The required rigidity of the structure or unit is then derived from its shape rather than from the material; the strength of the structure is, of course, only a function of the strength of the material. Much of the pioneering work in the use of G.R.P. as a structural material was undertaken at Massachusetts Institute of Technology, under Professor A. Dietz and two of his best known books on the subject are given in the bibliography.

Surface structures such as domes, shells, hyperbolic paraboloids and folded plate systems show convincingly that stiffness is primarily a function of the geometry of the structure and these depend only on the strength properties of the material of which they are made. It is significant that all recent developments in plastics structures are curvilinear and are typical examples of stressed skin construction in which the skin not only forms the enclosure, but also contributes substantially towards carrying the external loads. There are various configurations possible for folded plate structures manufactured from prefabricated plastics sandwich panels. This type of structure is attractive to the manufacturer because, being composed of flat panels, moulds may be relatively simple.

Glass reinforced plastics laminates may be used for semi-structural applications; these include the construction of load bearing and infill panels. Corrugated and folded sheets of thermosetting materials reinforced with glass fibres have found a wide field of application.

Glass reinforced plastics laminates may also be used to form space grids.

These structures may be defined as three-dimensional developments of a system of intersecting horizontal trusses. Space grids consist of top and bottom members that are interconnected by vertical or inclined members. Vertical members are not often used and a double layer grid may be considered as consisting of square base pyramids in which the spaces are interconnected by members of glass reinforced plastics. Basically, there are two main types of double layer grids. These are:

(a) lattice grids consisting of intersecting vertical lattice girders;
(b) true space grids consisting of a combination of tetrahedra, octahedra or inverted pyramids having square, pentagonal or hexagonal bases.

This chapter describes some of the conventional as well as the unconventional forms into which the glass reinforced plastics composites may be moulded or constructed; the necessity for these shapes will be made clear by mathematical analysis in Chapter 8.

The structural shapes may be placed under one of the three following geometric forms:

(a) folded plates;
(b) synclastic and anticlastic shells;
(c) skeletal structures incorporating folded plates in the compression region of the construction.

Types of load bearing and infill panels will also be discussed.

7.2 Structural forms for use with G.R.P. panels and structures

The folded plate system is probably the most popular form used with G.R.P. components at present. This system may be subdivided into:

(a) prismatic;
(b) pyramidal;
(c) prismoidal;
(d) composite folded plates;
(e) shell structures.

7.2.1 Prismatic structural forms

A prismatic structure consists of rectangular plates supported on rigid end diaphragms. The most common form of this construction is illustrated in Figure 7.1, and it may be seen that this form has a basic repetitive beam unit which is assumed to span between the end diaphragms.

7.2.2 Pyramidal structural forms

A pyramid may have a regular polygon as its base; these polygons include a

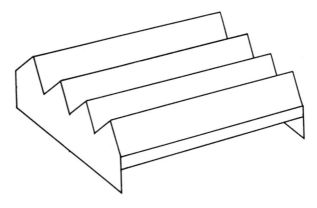

Figure 7.1 A common form of prismatic structure

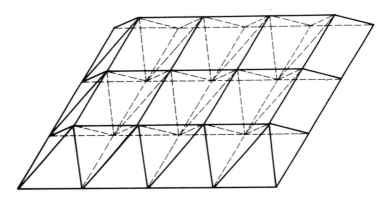

Figure 7.2 Modular pyramidal roofing on a square grid frame

triangle through to a circle. In designing a G.R.P. structure of this form, it would be necessary to incorporate a deep pyramidal form.

Plates forming the walls of a thin skin G.R.P. pyramid buckle under a comparatively low load intensity, and any increase of load will then be distributed to its support via the folds in the unit. Because of this premature buckling, large pyramids are not normally used as a structural form; the smaller pyramidal modules known as 'modular cladding' are preferred and these may have a variety of geometries and are supported by a grid frame. These forms of structures may be used for creating unusual roof shapes or for other special constructions. Figure 7.2 shows a modular pyramidal roofing on a square grid frame. Figure 7.3 is an actual example of this form of construction. Ball and Mullen[7.1], in a study of the overall cost/performance

Figure 7.3 Covent Garden Flower Market at Nine Elms, London. The roof which covers 1 ha is made of 924 G.R.P. mouldings each measuring 3.5 × 3.5 m. The mouldings were made for Mickleover Ltd by Armshire Reinforced Plastics with Scott Bader's Crystic polyester resin

for various forms of G.R.P. roofing, point out that the modular pyramidal roofing is likely to be more expensive than prismatic roofing, even after optimization of the design and production costs.

The structural systems consisting of three-dimensional space grids are well suited to prefabrication as they can be composed of a large number of relatively small elements. Together they form a stable and stiff structure which is able to withstand high concentrated loads. Local damage will seldom lead to total collapse. Makowski (see bibliography), some thirteen years ago, initiated research in which some of the skeletal members were replaced by the pyramidal units mentioned above and earlier examples of this form of construction show the pyramidal units connected at their apices and bases to two way skeletal grids; this form of construction is shown in Figure 7.4. The pyramids, which may be either square or hexagonal in plan, are made from G.R.P. and the skeletal grids from either steel, aluminium or undirectional glass reinforced polyesters.

One of the main advantages of the stressed skin space grids over the skeleton type double-layer grids lies in the fact that the former structure requires no separate roof decking as the stressed skins themselves act not only

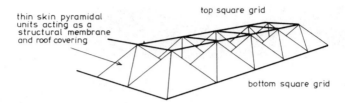

top square grid

thin skin pyramidal
units acting as a
structural membrane
and roof covering

bottom square grid

Figure 7.4 Pyramidal G.R.P. units connected at their apices and bases to two
way skeletal grids

Figure 7.5 A canopy at Arnham, Holland. The system consists of 40 pyramids,
each 1 m × 1 m at the base and having a height of 0.5 m. The apices are
interconnected by box shaped, foam filled G.R.P. ties with a filament wound
glass reinforcement in the longitudinal direction. Pyramids were manufactured by Labucon, Holland. (*Photograph by courtesy of P. Huybers, Delft
University of Technology*)

as a load bearing part of the structure but as a roof covering as well; the roof
decking in the latter structural form does not usually contribute to the overall
stiffness of the system.

Although the thin sheets of the pyramids will show unstable behaviour at
low stress levels, especially under compression and shear, at higher loads
when the stresses are concentrated around the fold lines, the overall stiffness
of the structure is high due to the supporting action of the top and bottom
grids. An actual example of these early structures is shown in Figures 7.5 and
7.6.

The method of stress analysis of such structures is given in section 8.4 and is
based upon the assumption that part of the continuum may be ignored and

Figure 7.6 Details of joints of the canopy at Arnham (*Photography by courtesy of P. Huybers, Delft University of Technology*)

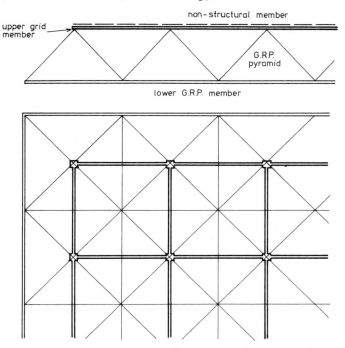

Figure 7.7 Double layer grids consisting of pyramidal units of square base; these units form the core and act as web membranes. The apices of the pyramids are connected by tubes of another material

that the folds of the units will then act as struts with effective areas determined from deflection studies. Gilkie [7.2] used and extended this method to thin walled pyramidal studies. Robak [7.3] undertook studies on double layer grids consisting of pyramidal units of square base; these units formed the core and acted as the web members. The apices of the pyramids were connected by tubes forming a skeletal grid, as shown in Figure 7.7. He demonstrated that G.R.P. pyramidal roofs are suitable for spans up to 25 m or to 35 m, depending on whether a single curved or double curved system is being used. Robak extended the above work when he investigated a G.R.P. roof system of two tiers of plastics pyramids jointed along a horizontal plane with the upper spaces filled with foamed plastics.

A further system which makes use of the pyramidal unit consists of bonding the base of truncated pyramids, which are themselves bonded together at the points of truncation, to the continuum membranes. In this case the pyramids form the core of the sandwich construction as shown in Figure 7.8.

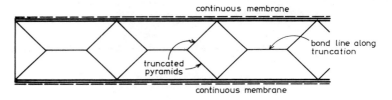

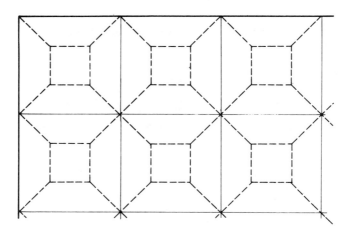

Figure 7.8 Truncated pyramids forming core of sandwich construction

Figure 7.9 Prototype G.R.P. school classroom erected at Fulwood, Lancs. under the direction of the Architect to Lancashire County Council. This is part of a proposal for a complete school system in G.R.P. (*Photograph by courtesy of the Architect to Lancashire County Council*)

The variety of possible layouts which can be obtained in stressed skin space grids is very wide and provides many fascinating patterns which are architecturally exciting.

The 'all plastics' structure is a variation of the above systems in which the pyramidal units are connected together to form the complete structure; the shape is generally structurally rigid. A typical example of this form is shown in Figure 7.9. The icosahedron geometric shape, which was chosen to form the prototype classroom, was constructed from a series of identical pyramids manufactured from the same mould. The whole structure may be defined as a continuous space structure.

7.2.3 Prismoidal structures

Prismoidal structures may be defined as an intermediate form of construction between a prismatic and a single pyramid structure. They consist of a prismatic structure whose end plates are inclined rather than vertical. Born [7.4] has given a number of examples of prismoidal structures, one of which is shown in Figure 7.10.

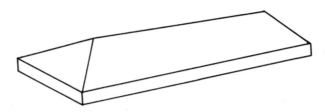

Figure 7.10 Prismoidal folded plate roof

7.2.4 Composite folded plate structures

This type of construction relies upon the folded plate action of the individual units being integrated with the elastic behaviour of the structure, with the result that its stiffness is increased in the longitudinal as well as the transverse direction. Line diagrams of composite folded plate structures are given in Figure 7.11. An investigation of a full size prototype, manufactured from a sandwich construction and made from a basic unit of diamond shape, was undertaken by Benjamin and Makowski [7.5]; the configuration is shown in Figure 7.11c. A mathematical method of analysis was developed from which designers are able to determine the stress distribution in such a system with a fair degree of accuracy; this method is discussed in section 8.3.

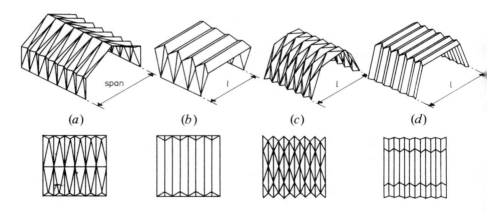

(a) (b) (c) (d)

Figure 7.11 Various forms of folded plate structures (after Makowski)

Huybers [7.6] has designed and tested a cellular barrel vault manufactured from prefabricated G.R.P. tetrahedrons; the two halves of the diamond tetrahedron are folded at right angles to one another; Figure 7.12 shows a test model of the barrel vault. There are a number of practical examples of composite folded plate structures and Figure 7.13 shows one built by Anmac Ltd. Nottingham.

Figure 7.12 Barrel vault model constructed from a number of tetrahedral units and vacuum formed from Sintilon. (*Photograph by courtesy of Structural Plastics Research Unit, University of Surrey*)

Figure 7.13 Interior view of sports hall used by Essex County Council at Braintree College of Further Education. Structure designed and fabricated by Anmac Ltd, Nottingham (*Photograph by courtesy of Anmac Ltd*)

7.3 Shell structures

Shell structures may be divided broadly into two basic types:

(a) singly curved shells;
(b) doubly curved shells.

7.3.1 Singly curved shells

The singly curved shell has zero curvature in one principal direction and at right angles to this the curvature at all points on the structure is of the same sign. The most familiar example of this type of shell is the barrel vault of circular cross-section which is usually supported on rigid end diaphragms; a line diagram of a barrel vault is shown in Figure 7.14. Figure 7.15 is a practical

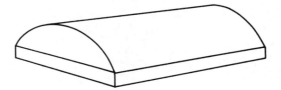

Figure 7.14 Barrel vault

Figure 7.15(a) Single skin elements used by R. Piano to form a singly curved shell manufactured from G.R.P. The elements are 2.60 m long with vertical flanges along the edges. The building is a 10.0 m span sulphur factory

Figure 7.15(b) Internal view of sulphur factory

example of a singly curved shell manufactured from G.R.P. and designed by R. Piano (Italian architect). The client required a light and transportable structure to cover a sulphur ore processing plant that could be moved from one location to another over the extraction field. In addition, the roofing material had to be resistant to chemical reaction with the sulphur vapours. A structure manufactured from G.R.P. material completely satisfied the two main requirements.

7.3.2 Doubly curved shells

The double curved shells may be of two types:

(a) Synclastic shells in which the two principal curvatures are of the same sign (positive Gaussian curvature); as shown in Figure 7.16. The simplest example of this shell is the dome in which the two curvatures are of equal magnitudes although this is not a necessary condition (e.g. the elliptic paraboloid). For G.R.P. shells of positive curvature the edge effects tend to damp rapidly and as a rule are confined to a narrow zone within this region. It follows that membrane theory may be used to determine the stresses over the structure.

(b) Anticlastic shells in which the principal curvatures are of different sign; this is illustrated in Figure 7.17 (the surfaces are of negative Gaussian curvature). For G.R.P. shells of negative curvature the boundary effects are much more significant than those for positive curvature and

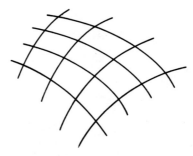

Figure 7.16 Synclastic surface from which a synclastic shell is made

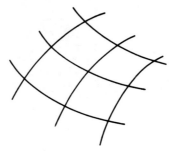

Figure 7.17 Anticlastic surface from which an anticlastic shell is made

therefore the design of a stiffening edge element must be carefully considered.

The most efficient single skin construction for G.R.P. is the one in which the material is mainly in tension; difficulties due to overall instability may be encountered in the design of a structure of synclastic geometric form in which the material is principally in compression. As the principal curvatures of anticlastic surfaces are of opposite sign, the buckling of thin skins of G.R.P. in one direction is virtually eliminated by the tension in the other direction.

One of the most common anticlastic surfaces is generated by the motion of straight lines along two skew lines; this is shown in Figure 7.18 where the geometric shape is known as a hyperbolic paraboloid or hypa. These shapes may also be obtained by translating a negative parabola on a positive one. Hypas have considerable resistance to buckling for the reasons given in the previous paragraph.

Zerning[7.7] has given a procedure for the fabrication of hypa moulds in which use is made of the 'cocoon-spinning' technique in which steel wires are tensioned along the straight lines of the hypa surface formed by timber formwork. These cables may be as much as 800 mm spacings. On top of this network a quick drying vinyl latex is sprayed and this becomes tensioned

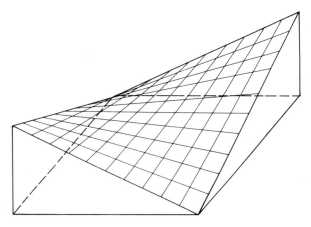

Figure 7.18 The formation of a hyperbolic paraboloid

during setting as shrinkage takes place. Figure 7.19 shows a photograph of a model developed and manufactured by Zerning. Figure 7.20 shows a swimming pool at Aberdeen consisting of hypa shell composites; it was designed by R.P. Structures Ltd.

Under a uniformly distributed load a hypa structure is under membrane

Figure 7.19 Dome manufactured from hyperbolic paraboloid units

K

Figure 7.20 Swimming pool building near Aberdeen. Dimensions 15 m long and 7.5 m wide; 4 m high. Constructed from 50 translucent hyperbolic paraboloid modules each measuring 4 × 1.5 m. Moulded by Aberglen Glass Fibre with Scott Bader's Crystic polyester resin (*Photograph by courtesy of Scott Bader Ltd*)

stresses; when, however, a non-uniform load or point load is applied to it, bending moments arise within the material; these need to be taken into account when designing.

7.4 Skeletal structures manufactured from G.R.P. and C.F.R.P.

With the pultrusion technique being utilized to produce G.R.P. composites of very high fibre content (e.g. 70%) and consequently high strength, it is conceivable that skeletal structures will be manufactured from this material within the foreseeable future. An experimental skeletal structure made from C.F.R.P. is shown in Figure 7.21.

7.5 Miscellaneous G.R.P. structures

Probably the most spectacular use of G.R.P. as main load bearing units has been provided by radomes. These installations house delicate radar equipment and the requirements for the material forming its cover is that the

Figure 7.21 An experimental skeletal structure, manufactured from C.F.R.P., under test in the Structural Plastics Research Unit, University of Surrey

structure must be transparent to electromagnetic waves. G.R.P. meets this requirement. Many radomes have been built from prefabricated components in G.R.P. Recently the Air Traffic Control Systems in the U.S.A. have experienced disturbance due to undesired radar reflections and beacon target splitting; the prime source of interference has been traced to Metallic Radar Micro Wave Link Towers which are situated close to the radar antenna. In order to minimize this interference a 30.5 m G.R.P. tower system has been inserted into the steel tower at the Chicago McCook Ill. air route surveillance radar at a height where the radar sweep would normally penetrate. The tower section was manufactured from G.R.P. members and designed to stringent requirements. The members were hollow oval shapes and incorporated the least vertical members of the smallest possible size. Generally, G.R.P. would not be considered for use in tall slender structures as a primary framing material. However, there are certain projects which negate the feasible use of traditional materials and in these cases the use of G.R.P. is not only a practical necessity but proves to be structurally efficient.

The entire roof of Morpeth School, Stepney, London, U.K., built in 1973 is in G.R.P. and is shown in Figure 7.22. Clear spans of up to 17 m are carried entirely in the G.R.P. trough sections. In the sides of the troughs are translucent panels which are a continuation of the stress carrying structure. In this application G.R.P. is being used to perform many building component functions simultaneously.

Figure 7.22 Morpeth School, Stepney, London. Consulting engineers: Nachshen, Crofts and Leggatt. G.R.P. manufacturers: Anmac Ltd.

Figure 7.23 The G.R.P. chimney at Hendon, London. Consulting engineers: Nachshen, Crofts and Leggatt. G.R.P. manufacturers: R. Graydon Ltd, Clients: Metropolitan Police.

The G.R.P. chimney at Hendon was built in 1970 and is shown in Figure 7.23. It is 37 m high and carries the flues of four large boilers. In order to give an efficient and stable structural form, full advantage was made of the positioning of the four separate flue shafts. The stress bearing G.R.P. tubes were formed by filament winding; the exposed G.R.P. face was the open mould free surface. The lining of the flue shaft is of steel.

7.6 Load bearing and infill panels

The principal outlets for reinforced plastics into the construction industry have been limited to categories of buildings where their use has not contravened the requisite national and local government building regulations. Over the last 20 years the industrialized building systems have been gradually taking the place of the traditional ones. With this change and with new technologies and materials, it is not surprising to find that the potential of glass reinforced plastics was considered and then exploited firstly as cladding panels before graduating to complete structures. The trend in the building industry at present is to use this material as cladding and load bearing panels and probably the majority of plastics is used in the industry in this way when increased productivity has to be achieved by a small static labour force.

Infill panels may have to carry superimposed loads such as wind, snow and maintenance and will therefore have to be designed accordingly. Probably the most impressive example to date in which G.R.P. has been used as a load bearing infill panel is the Mondial House building and a photograph of this is shown in Figure 7.24. Jointing details and descriptions of the panels are given

Figure 7.24 Mondial Internal Telephone Service, Central London. Clad with G.R.P. panels moulded by Anmac Ltd, Nottingham with Scott Bader's Crystic polyester resin. (*Photograph by courtesy of Scott Bader Ltd.*)

Figure 7.25 Part of H.M.S. Raleigh Redevelopment Scheme

Figure 7.26 The American Express Building (*Photograph by courtesy of Brensal Plastics*)

in Chapter 9, example 9.13. Glass reinforced plastics in panel form is also frequently used in offices, shops and hospitals. An example of this is demonstrated in Figure 7.25 which shows a part of the H.M.S. Raleigh redevelopment scheme.

The American Express Building (shown in Figure 7.26) in Brighton was completed in 1977. It is an interesting, practical and unique example incorporating all the advantages which can result from employing G.R.P. for building purposes. It is understood to be the first time that a G.R.P. element has been used in a semi-structural sense for a large prestige building. The white G.R.P. panels are, in effect beams and these span 7.2 m from column centre to column centre with several larger panels spanning 12.5 m. The 2 m high laminated glazing system, tinted blue, is supported on these G.R.P. panels to effectively divorce the building envelope from the building structure, thus obviating problems of building, tolerance, fit and deflections of the pre-cambered floor slabs. The G.R.P. panels had, therefore, to accept not only direct wind loads, but also both the weight of the windows and the wind and other loads imposed on them. Details of the panels of this structure are given in Chapter 9, Example 9.14.

7.7 Filament wound G.R.P. pipes

The technique of filament winding may be used for pipelines which are installed below ground for the conveyance of sewage, surface water or other effluents under gravity or pressure. Filament wound pipes consist essentially of thermosetting synthetic resins selected according to the corrosion duty required. They are reinforced directionally with E glass fibre rovings to give the required levels of strength both circumferentially and longitudinally. In order to achieve circumferential strength the continuous rovings are wound approximately at right angles to the longitudinal axis of the pipe and the axial strength is achieved by laying fibres parallel to this longitudinal axis. The same effect may also be achieved by winding the continuous rovings helically at an angle selected to give the stress resistance in both directions. In addition, there is a resin rich outer surface and the inner surface of the pipe is reinforced with synthetic fibres providing a high degree of resistance to abrasion and corrosion.

References

7.1. BALL, P.E. & MULLEN, D.C., 'G.R.P. roofing—cost performance study', Paper Number 52, International Symposium on Roofs and Roofing, Brighton, England, September, 1974.
7.2. GILKIE, R.C., 'Pyramids in lightweight roof systems', Ph.D. thesis, University of Surrey, 1967.

7.3. ROBAK, D., 'The structural use of plastics pyramids in double-layer space grids', in *Proceedings of International Conference on Space Structures*, Blackwell Scientific Publications Ltd., London, 1967, pp. 737–752.

7.4. BORN, J., *Hipped Plate Structures*, Crosby Lockwood, London, 1962.

7.5. BENJAMIN, B.S. & MAKOWSKI, Z.S., 'The analysis of folded plate structures in plastics', in *Plastics in Building Structures*, Pergamon Press, 1965, pp. 149–163.

7.6. HUYBERS, P., 'See-through structuring', Dissertation for doctorate, Delft University of Technology, Holland, 1973.

7.7 ZERNING, J., M.Phil. thesis, University of Surrey, 1970.

Bibliography

1. ALLEN, H.G., 'Characteristics of plastics and their use in structures', *Proc. Inst. Civil Engineers*, March, 1967, pp. 533–555.

2. Du CHATEAU, S., 'Couvertures de grandes portées en matière synthétique', *Neuf*, Sept. -Oct. 1969, pp. 16–25.

3. Du CHATEAU, S., 'Industrialisatie van ruimterakwerken', *Industrieel Bouwen, J.* 1970, pp. 587–595.

4. Du CHATEAU, S., 'Die Kunststoffanwendung bei weitgespannten Uberdachungen', *Plasticonstruction*, **2**, 1972, pp. 53–59.

5. DAVIES, R.M., *Plastics in Building Construction*, Blackie and Son, 1965.

6. DIETZ, A.G.H., *Engineering Laminates*, Wiley, 1948.

7. DIETZ, A.G.H., *Engineering Laminates*, M.I.T. Press, 1969.

8. DIETZ, A.G.H., 'Plastics in load-bearing constructions', Int. Sym. on Plastics in Building, April, 1970, Rotterdam.

9. FEIERBACH, W., 'Kunststoffhaus fg. 2000' in *The Use of Plastics for Load Bearing and Infil Panels*, (ed.) HOLLAWAY, L., Manning Rapley Publishing, 1976, pp. 213–214.

10. GAMSKI, K., 'Plastics in structures—European practice', ASCE Nat. Str. Eng. Meeting, April, 1972, Cleveland, Ohio.

11. GIBBS & COX LTD., *Marine Design Manual for Fibre Glass Reinforced Plastics*, McGraw-Hill, 1960.

12. GROOT, C.J.W.P., '"Play-dome" te Amsterdam', *Plastica*, April, 1971.

13. HOLLIDAY, L. (ed), *Composite Materials*, Elsevier, 1966.

14. JONES, R.M., *Mechanics of Composite Materials*, McGraw-Hill, 1975.

15. KELLY, A. *et al.*, *Composite Materials*, Iliffe, 1966.

16. KLUNDER, J., *Silenka Service Manual on Glassfibres*, Hoogezand, 1970.

17. MAKOWSKI, Z.S., 'Plastics as components in dome and roof structures', *Transactions and J. of the Plastics Inst.*, Feb., 1960, pp. 26–29.

18. MAKOWSKI, Z.S., 'Structural use of plastics in stressed-skin construction', *Applied Plastics*, Feb., 1963, pp. 47–52.

19. MAKOWSKI, Z.S., 'Structural Plastics', Proc. of Conf. 'Industrialized building and the Structural engineer', Organized by the Inst. of Str. Engrs., London, 1966, pp. 194–205.

20. MAKOWSKI, Z.S., 'Recent developments in structural applications of plastics materials', Proc. 26th Annual Technical Conference of Society of Plastics Engineers, New York, May, 1968, pp. 359–368.

21. MAKOWSKI, Z.S., 'The structural applications of plastics', Proc. of the Symposium: Plastics and Building Tomorrow, EFTA Plastics Association, Copenhagen, 1972, pp. 11–19.

22. McCRUM, N.G., *Review of the Science of Fibre Reinforced Plastics*, HMSO, 1971.

23. OBERBACH, K., 'Das Verhalten von Kunststoffen bei Kurzzeitiger und langzeitiger Beanspruchung und Kernfunktionen', *Zeitschrift für Werkstofftechnik*, June, 1971, pp. 281–291.

24. OLEESKY, S.S. *et al., SPI Handbook of Technology and Engineering of Reinforced Plastics/Composites,* Van Nostrand Reinhold, 1973.
25. PARKYN, B. (ed.), *Glass Reinforced Plastics.* Iliffe, 1970.
26. PIANO, R. 'L'impiego strutturele delle materie plastiche', Plast 68, Milano, 7/10 Oct., 1968.
27. PIANO, R., 'Experimental project of shell structures', *Casabella,* 335, April, 1969.
28. 'Swimming pool cover erected in two days', *Reinforced Plastics,* June, 1969, pp. 268.
29. ROSATO, D.V. & GROVE, C.S., *Filament Winding,* Wiley, 1964.
30. SAECHTLING, H.J. *Bauen mit Kunststoffen,* Carl Hanser Verlag,Munchen, 1973.
31. SCHEICHENBAUER, M., 'Construire en matière plastique', *Plastiques Bâtiment,* Dec, 1969, pp. 44–51.
32. SCHEICHENBAUER, M., Contribution in *The Use of Plastics for Load Bearing and Infil Panels,* (ed.) HOLLAWAY, L., Manning Rapley Publishing 1976, pp. 212.
33. SCHWABE, A., 'Kunststoff-Uberdachungen', *Plasticonstruction,* Jan, 1972, pp. 33–52.
34. SKEIST, I., *Plastics in Building,* Reinhold, New York, 1966.
35. SPENCER, A.J.M., *Deformations of Fibre Reinforced Materials,* Oxford University Press, 1972.
36. VINSON, J.R. & CHOU, T.W., *Composite Materials and Their Use in Structures,* Applied Science, 1975.
37. WENDT, F.W. *et al.* (eds.), *Mechanics of Composite Materials,* Pergamon, 1970.

8 Analyses and design methods for reinforced plastics structures

This Chapter is divided into two parts; the first one deals with design methods for complete G.R.P. structures and the second with component parts of structures.

PART I DESIGN METHODS FOR COMPLETE G.R.P. STRUCTURES

8.1 Introduction

It has been shown in Chapter 7 that structures and components of structures which are manufactured from G.R.P. materials must be folded or shaped so that the stiffness of the overall structure is derived from its shape rather than from the material. This present chapter deals with the analysis of such structures. There are various techniques for the analysis of folded plate and stressed skin structures and these will be discussed and summarized. They are applicable only to structures of particular configurations and any variation or modification which has to be made to the analysis because of the peculiarities of G.R.P. will be discussed. The joints of folded plate structures are in most cases rigid (i.e. they are capable of transferring transverse moments) but on occasions they may also be hinged.

The carrying capacity of folded plate structures is derived from statical depth d of the whole assembly (Figure 8.1), whilst the thickness t of the plate element is in most cases very small; in sandwich construction, however, the thickness could be large.

8.2 Method of analysis for simply supported prismatic folded plates

Simplified methods of analysis are constantly being sought but those which succeed generally suffer from severe limitations in their applicability to

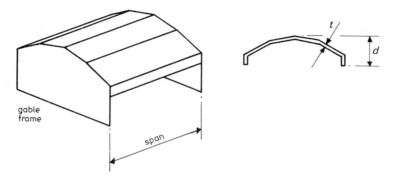

Figure 8.1 Prismatic folded plate structure

particular structural situations. The procedure set out in this section has the advantage that it does not require a solution by computer, but it is limited to one specific type of configuration.

Prismatic folded plate structures are generally considered to span longitudinally between rigid end diaphragms; this is usually referred to as the longitudinal plate action of the structure. In addition, the folded plates act as slabs which withstand transverse bending and shear forces; this is known as the transverse slab or plate action. The above structure is represented in Figures 8.1 and 9.1.

The method of analysis may be divided into three parts [8.1]:

(a) transverse slab analysis;
(b) longitudinal plate analysis;
(c) correction analysis.

The assumptions made in the analysis are:

(a) the material is elastic, isotropic and homogeneous;
(b) transverse slab action implies bending normal to the planes of the plates only;
(c) longitudinal action implies bending within the plane of the plate only;
(d) there is a linear variation of stress over the depth of the longitudinal beam;
(e) the plates have no torsional stiffness normal to their own planes;
(f) the diaphragms of the supports of the longitudinal beam are infinitely stiff in their own planes and completely flexible normal to their own planes;
(g) the distortions due to forces other than bending moments are negligible.

It should be stressed here that the method is applicable only to prismatic folded plate structures with rectangular plates; the solution of a composite folded plate structure would be difficult to solve by this method. (It is advisable to read the following sections in conjunction with example 9.1.)

8.2.1 The transverse slab analysis

All the externally applied loads are considered to be carried transversely by plate action. The plates act as continuous slabs spanning between unyielding supports at the folds. A moment distribution may be used to compute the slab moments, shears and joint reactions.

8.2.2 Longitudinal plate analysis

All the loads carried at the joints in the transverse slab analysis are considered transferred longitudinally to the end supporting members. These joint reactions are resolved into components in the planes of the plates of the prismatic structure, thus obtaining the coplanar loads which produce a state of plane stress in the plate. The resulting longitudinal stresses are determined on the assumption that each plate carries its load independently. Stresses at the extreme fibres of the assumed beams are not equal and this incompatibility will produce longitudinal shear stresses to equalize the edge stresses. The shear stresses can be calculated by the solution of a set of linear simultaneous equations based on the condition that the longitudinal strains of the edges of the plates adjacent to a joint are equal. However, an alternative method which consists of a distribution procedure, analogous to moment distribution and not requiring the setting up and solution of equations, is usually utilized; the application of this method is demonstrated in example 9.1. From the equalized edge stresses the individual plate deflection at mid span may be computed and the relative joint displacements are obtained by geometry. The relationship between plate deflections and relative joint displacements is given by:

$$\Delta_n = -\frac{\delta_{n-1}}{\sin \alpha_{n-1}} + \delta_n \left(\cot \alpha_{n-1} + \cot \alpha_n \right) - \frac{\delta_{n+1}}{\sin \alpha_n} \tag{8.1}$$

The verification of this formula is given in Appendix 8.1 and Figure 9.1.3 shows the geometrical relationship between plate deflection and relative joint displacement for the worked example 9.1. It is also shown in example 9.1 that a transverse distortion of the cross-section of the structure invariably results; this does not satisfy the assumption made in the analysis that all joints deflect equally. Although the above analysis satisfies statics there is an incompatibility between the displacements found from the longitudinal plate analysis and those assumed in the transverse slab analysis. In order to obtain compatibility a correction is applied to the above analysis.

8.2.3 Correction analysis

The correction analysis is designed to deal with the transverse distortion of the cross-section arising from the previous analysis. In order to reduce the required calculation to manageable proportions, it is assumed that the relative joint displacements vary longitudinally in a sine wave of period $2L$.

8.2.3.1 Transverse slab analysis

An arbitrary relative joint displacement Δ (conveniently taken as unity) is successively applied to each plate and the resulting fixed end moments determined. The fixed end moments will be determined from expression

$$(3EI\Delta)/h^2$$

or $$(6EI\Delta)/h^2$$

where the notation has its usual meaning.

By a moment distribution the transverse moments, shears and joint reactions are computed based upon the relative joint displacements at mid span; these reactions must be provided by the longitudinal plates.

8.2.3.2 Longitudinal plate analysis

The applied joint displacements are assumed to vary longitudinally in a sine wave [8.2] and the reactions and plate loads will also have a similar variation; the longitudinal moment at mid span of any plate will be:

$$M = \frac{PL^2}{\pi^2} \tag{8.2}$$

where P = the intensity of load at mid span.

The free edge stresses in the longitudinal plates are calculated from the equation.

$$\sigma_b = -\sigma_t = \frac{M}{Z} = \frac{PL^2}{\pi^2 Z} \tag{8.3}$$

where Z = section modulus

$$= \frac{1}{6}th^2$$

The deflection at the centre of the span will be given by

$$\delta = \frac{ML^2}{\pi^2 EI} \tag{8.4}$$

where $M = (\sigma_b - \sigma_t)\dfrac{I}{h}$

and hence $\delta = \dfrac{(\sigma_b - \sigma_t)}{Eh}\dfrac{L^2}{\pi^2}$ $\tag{8.5}$

where h = depth of beam.

The resulting stress incompatibilities are again removed by the application of a stress distribution method; plate deflections are then computed from equation (8.5).

The total centre span deflection of the longitudinal plates is the summation of the deflection obtained from the original analysis and the product of the above deflection obtained from equation (8.5) and the actual relative joint displacement $\Delta_\alpha, \Delta_{\alpha+1}$, etc. These total plate deflections may now be related to the actual relative joint displacements by geometry using equation (8.1) from which simultaneous equations are formed, the total number of which equals the number of restrained plates in the system. The true values of the stresses, moments and deflections due to the actual relative joint displacements are now computed, and the raw and corrected analyses are combined by the principle of superposition to give the final values of the various components.

8.2.4 Limitations of the method applied to prismatic G.R.P. structures

There is a severe limitation to the use of the above method for the analysis of G.R.P. prismatic folded plate structures, in that the deflection of the transverse plates must be small in comparison to the thickness of the plates. If sandwich construction is used, it is unlikely that the deflection will be large but if a chopped strand mat laminate is used there is a possibility that large deflections will occur. If this happens large deflection theory must be applied to all cases in which the composite is bent to a non-developable surface and where the stretching of the middle surface of the composite must be taken into account. This leads to non-linear equations and the problem becomes considerably more involved.

If the large deflection situation is likely to occur, Benjamin and Makowski [8.3] have suggested a general approach which is applicable to any shape or type of structure in single or in sandwich construction. This method is discussed in section 8.3.

8.3 Modified approach to the analysis of folded plate structures

Benjamin and Makowski [8.3] have suggested a method of analysis similar to the previous one in that a transverse slab and longitudinal plate system is considered; however, the analysis differs as follows:

(a) The transverse behaviour is considered by a complete slab analysis and not merely by a transverse strip analysis.

(b) If the deflections exceed half the thickness of the plate, large deflection theories are applied; otherwise small deflection theories are used.

The authors have stated that in the longitudinal direction the simplified methods are generally adequate, except that modifications should be made for particular cases such as:

(a) For plates having a span plate-width ratio of less than 1.5, the straight-line stress distribution is no longer valid and deep beam theory should be used.

(b) For plates where the plate-width to thickness ratio is fairly low (less than 100) and where the transverse loads, deflections and compressive stresses are not excessive, the entire width of the plate may be considered as effective when subjected to bending moments. When, however, this ratio is high (greater than 100) and when the transverse loads, deflections and compressive stresses in the plane of the plate are large, the buckled central portion of the plate does not contribute fully to its stiffness. In this case the compressive straight-line stress distribution will pass through successive stages as shown in Figure 8.2, gradually transferring load on to the stiff folded ridges of the structure.

(c) In the case of barrel vaults, the longitudinal behaviour is the portal-frame or arch action and the structure may be analyzed on its neutral axis by the usual methods of structural engineering.

In reference [8.3], the authors have discussed several folded plate structures, including the barrel vault; a discussion of this structure will be made here.

The transverse slab action for the barrel vault is determined by the analysis of a triangular slab under the external load normal to the surface. The edge fixing conditions of the slab would depend upon whether the folded plates were bonded together in the factory or bolted on site. As the deflection would invariably be of the order of the thickness of the slab, the analysis would be difficult. The authors have suggested that the arch action may be considered by assuming the arch to have a constant moment of inertia I, the inverse of which can be approximately determined by applying Simpson's rule to the fluctuating values of $1/I$ over the arch. The cross sectional area will be constant.

It may be shown that the largest transverse moments occur in the top horizontal units where the normal or transverse component of the vertical load is a maximum and therefore the loading on the units may be considered as point loads so that each panel supports one such load at its centre as shown in Figure 8.3. The point loads on the ridge and valley lines do not enter into the calculations for transverse behaviour, but have to be considered in the longitudinal analysis. The authors have found that when a long narrow plate has a point load applied only at the centre, the influence of the load does not extend into the corners of the plate but is limited as shown in Figure 8.3. The analysis will then be for a circular plate loaded at the centre and subjected to large deformations.

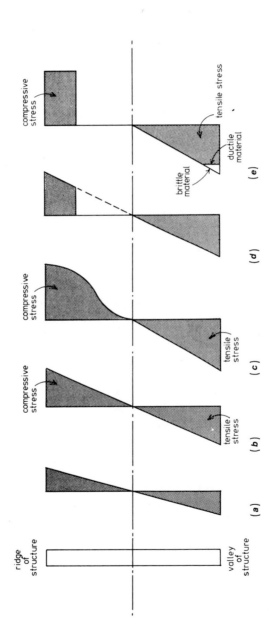

Figure 8.2 Stress-strain distribution for deep plates of folded plate structures: (a) strain diagram (for low values of applied load); (b) stress diagram (for low values of applied load); (c) possible stress distribution at higher values of applied loads; (d) final strain distribution; (e) possible final stress distribution

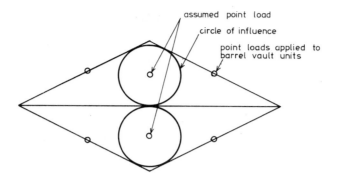

Figure 8.3 Developed plan of top horizontal unit of barrel vault (after ref. [8.3])

The deflected shape [8.4] may take the form:

$$w = w_0 \left[1 - \frac{r^2}{R^2} + \frac{2r^2}{R^2} \log \frac{r}{R} \right] \tag{8.6}$$

where w_0 = maximum deflection at the centre
      R = radius of the circle
      w = deflection at radius r.

The maximum deflection w_0 can be determined by the non-linear equation [8.4] :

$$\frac{w_0}{t} + A \left[\frac{w_0}{t} \right]^3 = \frac{BPR^2}{Et^4} \tag{8.7}$$

where t = thickness of plate
      P = point load
      E = modulus of elasticity of material.
      A and B are constants and are functions of the edge conditions of the plate and of the Poisson's ratio for the material.

Benjamin [8.5] has mathematically determined these constants as

    edge free   : $A = 0.194$; $B = 0.209$
    edge restrained: $A = 0.453$; $B = 0.209$

8.4 Skeletal configuration approach to the solution of continuum structures

Unlike the above, the following methods for structural analysis of G.R.P. folded plate structures depend upon the availability of a computer.
    A general solution, which gives stress analysis results [8.6] to composite

L

folded plate structures or to 'all plastics' structures manufactured from individual cells or units may be obtained by reducing the continuum structure to an equivalent skeletal one; the members of the latter structure are positioned at the folds of the former and the analysis is undertaken by matrix methods of structural analysis, such as the flexibility or the stiffness method. Reference [8.7] discusses these procedures and there are a number of programs available in most computer centre libraries for the solution of these methods. There are, however, three main difficulties when utilizing this procedure although these are not insuperable; these are:

(a) There is difficulty in evaluating the axial stiffness AE and the flexural stiffness EI for the equivalent skeletal members; some theoretical and experimental work on prismatic plates and angle sections has been undertaken [8.8]. As some of the continuum is ignored when calculating the areas of the skeletal members the overall deformations of the structures err on the side of safety.

(b) It is not possible to obtain directly the stresses within the continuum of the structure; to obtain these, a modified finite element approach has to be applied to the skeletal analysis results.

(c) A decision has to be made as to whether the skeletal structure should be treated as pin jointed or rigidly jointed.

When considering an individual unit of pyramidal form, as shown in Figure 8.4, it may be demonstrated that under an apex load of low magnitude the continuum of the pyramid will be stressed as shown. When, however, the applied load is raised to a certain value, the centre part of the continuum will buckle and any increase in load beyond this value will be supported by the folds. In structures of this form, therefore, it is advisable to increase the thickness of the material in the region of the folds above that of the centre part of the continuum; it may then be assumed that the load bearing part of the structure is derived only from the material in the folds. As a general guide this region may extend a distance of $16t$ (where t is the thickness of the plate) on either side of the ridge or valley lines of the folds.

If this procedure is adopted, the stresses in the various equivalent skeletal members derived from the axial forces and bending moments, may readily be computed from the results of the analysis, and an indication of the point at which the fold will buckle under a compressive load will be obtained. An estimate of the overall deflections of the structure will also be given from the analysis.

With this method, the structural loading data to the computer program may be specified as loads applied either at the nodal points or as internal forces to the members of the skeletal structure. Whereas the former may give satisfactory results for the overall behaviour, it does not take into account any

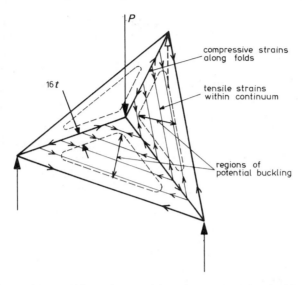

Figure 8.4 Strain condition of pyramid under an apex load of low magnitude

local bending of the equivalent skeletal members where the load is applied along their lengths; the latter case, however, does consider this factor.

The structure should generally be considered as a rigidly jointed one. However, it has been observed that with most structural configurations there is less than 5% difference in overall deflection of the structure between fully fixed and pin jointed conditions but the stresses in the material computed by these methods may vary by a greater amount due to the moment component. Generally the deflection of the overall structure is the design criteria, but it is also necessary to investigate thoroughly the compressive loads in the structural members, as these are generally the critical ones. The tensile stresses invariably have a very high factor of safety when the structure has been designed on limiting deflection.

The low stiffness combined with the high strength of G.R.P. composites makes buckling one of the prime considerations for design when utilizing the skeletal approach, and therefore the following section is devoted to a discussion of this problem.

8.5 Buckling of members under axial compressive forces

There are various types of buckling. The Euler buckling of a column is historically the earliest and occurs when a slender column is loaded in compression. The column is unable to carry any load greater than the buckling one. Plates and shells also buckle under in-plane loading and also develop 'local' buckling in the form of waves or ripples along the plate. This form of buckling, however, does not necessarily mean collapse of the structure as it can usually withstand loads beyond that required to initiate buckling. The post buckling behaviour of a plate must be ascertained if loads beyond the critical are anticipated.

If a load P is applied along the axis of a column, which is initially perfectly straight, it will remain straight until a critical value P_E of the load is reached; this is the Euler critical load. At this point the column becomes unstable and buckles.

If the compressive stress in every lamina of a ductile composite is less than the proportional limit of the material the critical load formula is

$$P_E = \frac{\pi^2 D}{(KL)^2} \tag{8.8}$$

where D = the flexural rigidity of the composite
$\quad\quad\ L$ = length of the member
$\quad\quad\ K$ = column's equivalent length factor.

If the column is under the action of a critical load and either

(a) the proportional limit of the material is exceeded and therefore the stress-strain relationship is non-linear; or
(b) the material, although elastic, does not follow Hooke's Law but has a compressive stress-strain relationship as shown in Figure 8.5,

then as the column buckles, bending will occur about the centroidal axis of the cross-section and there are corresponding increases and decreases in compressive strains towards the concave and convex sides, respectively, of the buckled column. Consequently, the instantaneous value of the modulus corresponding to the increases in the applied strain on the compressive side of the strut (the concave side) is the tangent value E_t; this quantity is a function of the applied stress. The modulus of elasticity on the opposite side of the composite, (the convex side) representing a decrease in strain will be defined by the initial modulus of elasticity value E. The conditions are analogous to a material having two different moduli, one for tension and the other for compression. Therefore, at the point of buckling, the equivalent modulus will be a function of both E and E_t and is usually referred to as the reduced modulus E_r; it will also be dependent upon the position of the lamina in the composite. The value of the reduced modulus must be formulated for each geometrical section. Timoshenko [8.9] has shown that for a rectangular section

$$E_r = 4 E E_t / (\sqrt{E} + \sqrt{E_t})^2$$

Thus a theory is available that meets the equilibrium condition of buckling at

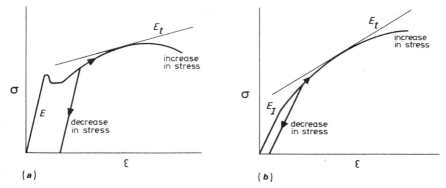

Figure 8.5 Stress-strain relationships for ductile and viscoelastic materials;
(a) ductile material; (b) viscoelastic material

E = modulus of elasticity (linear) $E = \sigma/\epsilon$
E_t = tangent modulus (non-linear) $E_t = d\sigma/d\epsilon$
E_r = initial tangent modulus

constant load, but Shanley [8.10] has shown that it is deficient because a perfectly straight column could not be expected to remain straight above the tangent-modulus load. He realized that bifurcation of equilibrium (see Appendix, p. 221) can occur at a load lower than that given by substituting $(D = I E_r)$ into equation (8.8); buckling is governed by the incremental flexural rigidity of the member. Therefore the load at which buckling becomes possible is the tangent modulus load P_t and is given by

$$P_t = \pi^2 D_t /(KL)^2 \tag{8.9}$$

where $D_t = E_t I$

The true failure load is difficult to obtain theoretically but it is known to be between the tangent modulus load P_t and the reduced modulus load P_r.

Glass reinforced plastics composites manufactured from polyester resins are assumed to be viscoelastic non-linear materials and therefore will approximately follow the above argument.

The controlling parameter in the above equations is the flexural stiffness; the effect of shearing forces on the value of the critical load has been ignored. With materials in which the resistance to shear effects is low, modification to the Euler equation must be made.

Timoshenko and Gere [8.11] have incorporated the effect of shear in equation (8.10).

Buckling load for a pin ended column $P_s = \dfrac{P_E}{1 + (nP_E/AG)}$ \hfill (8.10)

where P_E = Euler critical load given by equation (8.8)
 n = a numerical factor depending upon the shape of cross section
 = 1.11 for a circular cross section
 G = shear modulus.

From equation (8.10) it may be seen that by reducing the value of shear modulus the Euler critical load is reduced.

8.5.1 Flat prismatic plates under axial compressive forces

The application of the Euler critical load formula for long struts is acceptable when designing structures normally encountered in civil engineering. However, when short columns or struts manufactured from thin walled components (or plates) are considered the dominant criterion is plate buckling.

The problem of rectangular plates which are simply supported along their edges and under compressive loads has been examined by a number of investigators and the following paragraphs give the results of some of their theories.

Bryan [8.12] presented the analysis of the elastic critical stress for a rectangular plate simply supported along all edges and subjected to a uniform longitudinal compressive load. This stress for a long plate segment is determined by the plate width-thickness ratio b/t, by the restraint conditions along the longitudinal boundaries and by the elastic properties; it is given by:

$$\sigma_{cr} = \beta \, \sigma_e \tag{8.11}$$

where $\sigma_e = \dfrac{\pi^2 E}{12 (1 - \nu^2) (b/t)^2}$

t = thickness of plate or column

b = width of plate or column

a = length of plate or column

β = coefficient $\left[\dfrac{a}{m\,b} + \dfrac{m\,b}{a} \right]^2$

β determines a sequence of curves which corresponds to buckling in 1, 2, 3 half waves, depending upon the ratio a/b.

If the member cross section is composed of various connected elements, as in a stressed skin folded plate structure, a lower bound solution for the critical stress can be determined for each element by assuming a simple support condition if one edge of a member is attached to the edge of an adjacent plate element or a free condition if the plate is not attached to its neighbour. For any of the plate elements, the smallest value of the critical stress found is the lower

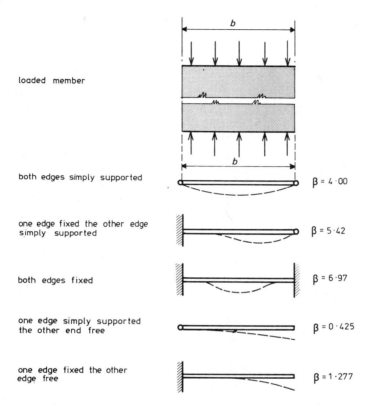

Figure 8.6 Approximate values of the buckling coefficients ß for different side fixing conditions of plates under uniform edge compression

bound of the critical stress for the cross section. Figure 8.6 gives values for specific edge supports which provide a good approximation when a/b is greater than unity which is generally the case in the equivalent skeletal approach; further information on the β coefficients may be found in reference [8.13].

A generalized expression for the critical buckling stress of a flat plate segment under uniform compression stress in either the elastic or inelastic range is given by Bleich [8.14] as:

$$\sigma_{cr} = \beta\, \sigma_e \qquad\qquad (8.12)$$

$$\sigma_e = \frac{\pi^2\, E\, \sqrt{\eta}}{12\,(1 - \nu^2)\,(b/t)^2}$$

$$\eta = E_t/E$$

$$\beta = \left[\frac{a}{mb} + \frac{mb}{a} \right]^2$$

As in equation (8.11) the value of β may be obtained from Figure 8.6.

Equation (8.12) is a modification to equation (8.11) in that it incorporates stresses which are either above the proportional limit or within the non-linear stress-strain relationship of the material. It is a conservative approximation to the solution of a complex problem. Although this equation was intended for materials with non-linear stress-strain properties it can be utilized to take into account residual stresses, geometrical imperfections and edge restraints by incorporating their effects into the tangent modulus of elasticity value E_t.

As examples of the use of the equations (8.11) and (8.12) it may be useful to refer to the paper by Korin and Hollaway [8.15] who have investigated experimentally and analytically the mode of buckling of some pultruded G.R.P. sections which could be used in a skeletal structural configuration.

8.5.2 Angle sections under axial compressive forces

It was stated previously (section 8.4) that continuum structures may be reduced to equivalent skeletal ones comprising angles whose leg lengths are $16t$. The designer must then satisfy himself that the Euler critical load for any angle is not exceeded. An alternative design is described below and this may be adopted to obtain a realistic value of the cross-section for the purpose of stability calculations.

When a flat rectangular plate supported along opposite edges is placed under a compressive load in a direction parallel to the edge supports, local instability can occur. The effect of such instability is to reduce the proportion of the compressive load carried by the centre region of the plate; the compressive stresses carried by the supported edges will continue to increase up to yielding. Von Karman et al. [8.16] investigated the buckling of primatic plates under axial compressive forces (see Figure 8.7) and showed that the ultimate compressive load P_u for a prismatic plate simply supported along the sides is:

$$P_u = \frac{\pi}{\sqrt{[3(1-v^2)]}} \sqrt{[(E\sigma)_c]}\ t^2 \tag{8.13}$$

and in general

$$P_u = C \sqrt{[(E\sigma)_c]}\ t^2 \tag{8.14}$$

where v = Poisson's ratio for the material of the plate

$(E\sigma)_c$ = product of the stress and modulus of elasticity in compression (i.e. product of the yield stress and the tangent modulus) or beyond the elastic limit it is necessary to replace the modulus of elasticity by the slope of the stress-strain graph (E_t) and to determine the maximum product of $(E_t\sigma)_c$

C = a constant depending upon various edge conditions.

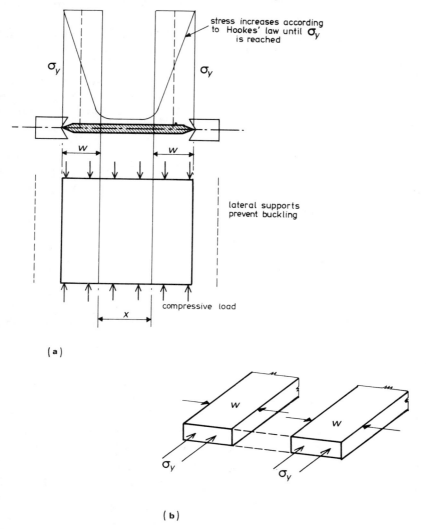

(a)

(b)

Figure 8.7 Thin plates under compressive forces (after von Karman[8.16]). (a) plate under compression (at x compressive stresses remain substantially constant as buckling deflections increase and may be neglected for thin sheets); (b) equivalent plate under compression (σ_y = yield stress of material)

The stress distribution in a buckled plate will be similar to that shown in Figure 8.7. It will be clear from equation (8.13) that the value of P_u is proportional to t^2 only and does not depend upon the length or width of the plate. Since the effect of a limited change of Poisson's ratio is minimal, equation (8.13) may be written in the form of equation (8.14).

As G.R.P. has no yield stress, the product $E\sigma$ is taken as the maximum value of the product of the stress at any point and the tangent modulus at that point.

Needham [8.17] considered methods of predicting the maximum average crippling stress (see Appendix 8.2 for definition of crippling stress) of formed sheet metal structural shapes in compression and proposed the following relationship for Vee, channel and square tube sections

$$\sigma_{cc} = \frac{C_c \sqrt{[(E\sigma_{pc})_c]}}{R_f^{0.75}} \qquad (8.15)$$

where $\quad \sigma_{cc}$ = average crippling stress for an individual angle element

σ_{pc} = 0.2% proof stress in compression

$\sigma_{cc} \not> \sigma_{pc}$

E = modulus of elasticity of the material in compression

R_f = buckling parameter = form ratio = $\dfrac{a+b}{2t}$ (see Figure 8.8)

C_c = coefficient that depends upon the degree of edge support from adjacent angle elements

= 0.316 for unsupported angles

= 0.342 for angle with one edge free

= 0.366 for angle with edges restrained.

It should be stressed that these coefficients were derived for aluminium alloy sheet material and are not directly applicable to G.R.P. materials.

Therefore the ultimate compressive load given by Needham is:

$$P_u = \sigma_{cc} \text{ total area}$$

the total area = $(a+b)t = 2 R_f t^2$ (see Figure 8.8)

$$\therefore P_u = \sigma_{cc} 2 R_f t^2 \qquad (8.16)$$

$$= 2 C_c R_f^{0.25} \sqrt{[(E\sigma_{pc})_c]} t^2 \qquad (8.17)$$

P_u is now a function of the plate width but any variation in a or b will affect the crushing strength of the section only marginally.

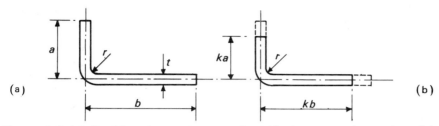

Figure 8.8 (a) Buckling factor $R_f = (a + b)/2t$ for an angle section. (b) reduction factor k for an angle section; reduced leg lengths ka and kb

In the above equation $(E\sigma_{pc})_c$ is equivalent to the product of the modulus of elasticity and the yield stress of the material, $(E\sigma)_c$ in Von Karman's equation, when ductile materials are being investigated with well defined yield points. When, however, materials are utilized which do not have a well defined yield point the two values may vary considerably. Frost [8.18] has suggested using the product $(E_t\sigma)_c$ defined in equation (8.14) in place of $(E\sigma_{pc})_c$ in equation (8.15) provided the values of C_c are replaced by C_c^1 as:

0.393 for unsupported angles
0.425 for angle with one edge free
0.455 for angle with edges restrained

This enables a wider use of equation (8.15) to cover such materials as G.R.P. As before the ultimate compressive load is:

$$P_u = \sigma_{cc} A = \sigma_{pc} k A \qquad (8.18)$$

when the radius of the bend is small (i.e. $r < 3t$)
where A = area of compression flange
k = leg reduction factor (see Figure 8.8)

from which $k = \dfrac{\sigma_{cc}}{\sigma_{pc}}$

Figure 8.9 shows the relationship, for an angle section of any material, between the dimensionless quantities $\sigma_{cc}/\sqrt{[E_t\sigma)_c]}$ and the form ratio R_f as given by Frost from the modified equation:

$$\sigma_{cc} = C_c^1 \sqrt{[(E_t\sigma)_c]} \ R_f^{-0.75} \qquad (8.19)$$

For angle sections the form ratio can be calculated and the value $\sigma_{cc}/\sqrt{[(E_t\sigma)_c]}$ can be found from the graph. As the value of $(E_t\sigma)_c$ is known, the value of the crippling stress σ_{cc} may be determined. Also the reduction factor may be determined from equation (8.18).

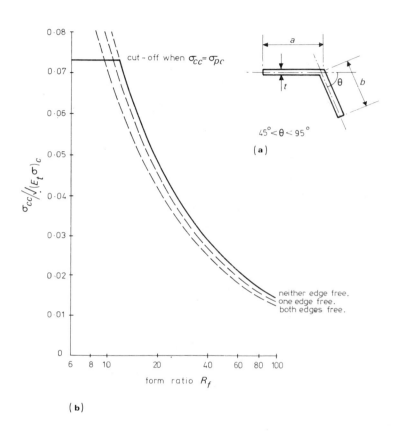

(b)

Figure 8.9 Relationship between $\sigma_{cc}/\sqrt{(E_t\sigma)_c}$ and form ratio R for angle section. (Based on Figure 6 of ref. 8.18)

8.5.3 Prismatic folded plates under combined compressive forces and bending moments

Figure 8.10 shows a part section of a member under both axial and bending stresses. Figure 8.10b shows the reduced section (the reduction clearly is in the compressive region only) where the leg lengths $a/2$ and y have been reduced by the leg reduction factor k.

An assumption made in deriving the strain diagrams in Figure 8.10c is that the distribution across the section is linear where:

ϵ_{pc} = strain at proof stress in compression (or equivalent)

ϵ_{pt} = strain in material at ultimate tensile failure.

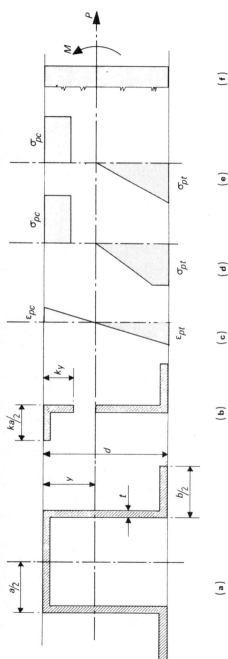

Figure 8.10 Top hat section under a compressive force and a bending moment: (a) top hat section; (b) reduced top hat section; (c) strain diagram; (d) stress diagram; (e) stress diagram; (f) part of section

Dimensions

a = width of compression flange
b = width of tension flange
d = depth of section
t = thickness of section

Stress diagram for ductile materials is shown in (d) and for brittle materials in (e) where for

Ductile materials

ε_{pc} = strain corresponding to compressive yield stress
ε_{pt} = limiting tensile strain
σ_{pc} = compressive yield stress
σ_{pt} = tensile yield stress

Brittle materials

ε_{pc} = strain at proof stress in compression
ε_{pt} = strain in material at ultimate tensile failure
σ_{pc} = proof stress in compression
σ_{pt} = ultimate tensile stress at failure

The stress distribution is obtained from the strain values. Knowing the effective areas of the section and the stress distribution, the following may be computed:

$$\text{the axial force} = \Sigma\,\sigma\,\delta A \qquad (8.20)$$

$$\text{the moment} \quad = \Sigma\,\sigma\,y\,\delta A \qquad (8.21)$$

The compressive load carrying capacity of angle sections is very dependent upon the angle θ of the two plates (shown in Figure 8.9a). Frost has shown that the optimum angle for maximum load carrying capacity lies within the range 45° and 95°; within this range the crushing strength remains constant to within 2%.

8.6 The finite element technique

A sophisticated analytical solution, which provides values of strains in and deflections of a loaded continuum structure, is the finite element technique [8.19]. This analysis requires the provision of a computer with the capability of a large store and some knowledge of the finite element technique. The technique assumes that the elastic continuum is approximated by an assemblage of finite elastic elements which are subjected to some of the constraints that exist in the continuum. Element stiffnesses are computed and combined into a master stiffness matrix which is utilized to solve the problem.

It is usual to satisfy the compatibility of displacements across the interface, but this will generally produce stresses in adjacent elements which are dissimilar and therefore do not satisfy the condition of local equilibrium at points along the interface. The overall equilibrium of the elements, however, is provided by the forces applied at the nodal points.

The finite element technique initially idealizes the structure by subdividing it into an assemblage of discrete elements. A degree of engineering judgement must be applied to the idealization process as the subsequent analysis and the validity of the solution depend upon this idealized structure. To achieve greater accuracy in regions of high stress and strain gradient the elements are reduced in size.

As with the skeletal approach there are a number of finite element computer programs available for general use in most computer unit libraries. However, it is questionable whether the economics of design of a G.R.P. structure warrants the use of one of these programs as it is generally more reliable to obtain the sizes of members in the structure by one of the previous methods, and then to undertake a full scale test on the completed structure. If, as a result of these tests, it is necessary to stiffen the structure or add extra members this may readily be undertaken on site with relative ease.

PART II DESIGN OF COMPONENT MEMBERS

8.7 Introduction

In the solution of a structural problem the analysis of the overall structure is
undertaken first and this is followed by the design of the component parts. In
the following sections, methods are suggested for the design of three specific
types of components; those discussed are:

   (a) chopped strand mat laminates (isotropic materials);
   (b) unidirectional and bidirectional rovings (orthotropic materials);
   (c) a laminate consisting of a combination of the above two.

8.7.1 The design of a chopped strand mat laminate in tension

A chopped strand mat laminate is generally assumed to be isotropic for design
purposes and the strain equations will be those given in section 2.6.2. The
relationship between the stresses in the fibre, the matrix and the composite has
been derived from the basic equation (2.1) given in Chapter 2. Rearrangement
of this equation gives equation (2.3) which is a relationship between values of
composite, matrix and fibre stresses; the composite stress can then be
determined when the values of the other two components are known. The
worked example 9.2 utilizes this expression.

8.7.2 The design of an unidirectional and bidirectional laminate

Unidirectional and bidirectional roving and glass fibre cloth composites are
assumed to be orthotropic and are governed by the strain equations given in
section 2.6.1; therefore, these components will have different ultimate
strengths at various angles to the warp direction and must be analyzed
accordingly.

The relationship for the ultimate tensile stresses in the lamina principal axis
(1,2) and the reference axis (x,y) at an angle θ with the principal axis can be
derived from the expression:

$$\frac{1}{\sigma_{xx}^2} = \frac{\cos^4 \theta}{\sigma_{11}^2} + \frac{\sin^4 \theta}{\sigma_{22}^2} + \frac{\sin^2 \theta \cos^2 \theta}{\sigma_s^2} \tag{8.22}$$

where σ_{xx} = ultimate tensile stress associated with the reference axis

   σ_{11} = ultimate tensile stress associated with the lamina principal axis
   (i.e. warp direction)

   σ_{22} = ultimate tensile stress associated with the lamina principal axis
   (i.e. weft direction)

   σ_s = ultimate shear stress at section considered.

The equivalent relationships for the elastic modulus and shear modulus are given in section 2.6.1.

For a bidirectional woven roving composite it is apparent that the maximum ultimate tensile stress will be in the warp and the weft directions; at any other angle to the warp direction, the ultimate tensile stress of the material will be lower.

To find the ultimate stress at an angle θ to the warp direction when the application of the load is in this direction, it is necessary to apply equation (8.22) twice, firstly, to obtain a value of σ_s when σ_{11}, σ_{22} and σ_{45} (the applied load will be at 45° to the lamina axis (1,2)) are known from experiment, and secondly to obtain the value σ_{xx} using the values of σ_{11}, σ_{22} and σ_s and knowing the direction of the line of application of the load relative to the lamina principal axis (see worked example 9.4).

When a composite is manufactured from say a chopped strand mat laminate (isotropic material) and a woven roving laminate (orthotropic material) and the applied load is at an angle θ to the warp direction of the woven roving laminate (see Figure 9.5.1) then the ultimate tensile stress in the direction θ must be calculated first. The equal strain criteria for the composite of chopped strand mat and woven roving at $\theta°$ to the warp direction is applied, and the ultimate failure load is then calculated on the ultimate stress of one of the laminates, consistent with the equal strain criteria, and the calculated stress of the other. The worked example 9.5 illustrates the procedure.

8.7.3 Design of a chopped strand mat or a woven roving laminate in compression

The design of a chopped strand mat or a woven roving laminate in compression is undertaken as described in section 8.5; the stress strain relationships of the laminates must be considered in the design. The worked examples 9.6 and 9.7 illustrate the design procedure.

8.8 Factors of safety with respect to stresses

The factor of safety must be carefully chosen and should be selected with due consideration given to the nature of the loading, the detailed design of the laminate and of the environmental conditions to which the structure will be exposed. Safety factors have varied from 2 for static short-term loads to 10 for repeated impact loads and indeed those on stresses are often higher than this latter value, due to deflection limits for overall civil engineering structures.

It is difficult to give any ruling for precise values and therefore it must be left to the experience of the designer to provide a sensible one.

Appendix 8.1

Consider member (n) of a prismatic folded plate structure, with joints i and j. The acute angle between members $(n-1)$ and (n) is α_{n-1} and between members n and $(n+1)$ is α_n. The cartesian coordinate axes are as shown.

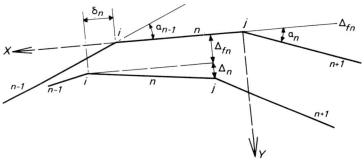

Figure 8.11

The assumption made in the derivation of the formula is that the length of the member does not change.

For member n

Displacement of joint i in the x-and y-directions are

$$\delta_{xi} = \delta_n$$

$$\delta_{yi} = \Delta_{fn}$$

Similarly for joint j

$$\delta_{xj} = \delta_n$$

$$\delta_{yj} = \Delta_{fn} + \Delta_n$$

Displacements at joint i transformed to system of coordinates of member $n-1$

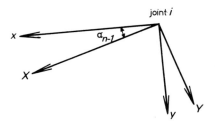

Figure 8.12

M

$$\begin{bmatrix} X \\ Y \end{bmatrix} = \begin{bmatrix} \cos \alpha_{n-1} & \sin \alpha_{n-1} \\ -\sin \alpha_{n-1} & \cos \alpha_{n-1} \end{bmatrix} \begin{bmatrix} x \\ y \end{bmatrix}$$

i.e. $$\begin{bmatrix} (\delta_{xi})_{n-1} \\ (\delta_{yi})_{n-1} \end{bmatrix} = \begin{bmatrix} \cos \alpha_{n-1} & \sin \alpha_{n-1} \\ -\sin \alpha_{n-1} & \cos \alpha_{n-1} \end{bmatrix} \begin{bmatrix} \delta_n \\ \Delta_{fn} \end{bmatrix}$$

$$(\delta_{xi})_{n-1} = \delta_{n-1} = \delta_n \cos \alpha_{n-1} + \Delta_{fn} \sin \alpha_{n-1}$$

$$\Delta_{fn} = \frac{\delta_{n-1}}{\sin \alpha_{n-1}} - \delta_n \cot \alpha_{n-1} \qquad (A.1)$$

Displacements at joint j transformed to system of coordinates $n + 1$

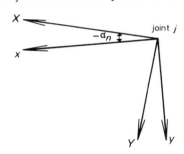

Figure 8.13

$$\begin{bmatrix} X \\ Y \end{bmatrix} = \begin{bmatrix} \cos \alpha_n & -\sin \alpha_n \\ \sin \alpha_n & \cos \alpha_n \end{bmatrix} \begin{bmatrix} x \\ y \end{bmatrix}$$

i.e. $$\begin{bmatrix} (\delta_{xj})_{n+1} \\ (\delta_{yj})_{n+1} \end{bmatrix} | = \begin{bmatrix} \cos \alpha_n & -\sin \alpha_n \\ \sin \alpha_n & \cos \alpha_n \end{bmatrix} \begin{bmatrix} \delta_n \\ \Delta_{fn} + \Delta_n \end{bmatrix}$$

$$(\delta_{xj})_{n+1} = \delta_{n+1} = \delta_n \cos \alpha_n - (\Delta_{fn} + \Delta_n) \sin \alpha_n$$

$$\Delta_n = -\Delta_{fn} + \delta_n \cot \alpha_n - \frac{\delta_{n+1}}{\sin \alpha_n} \qquad (A.2)$$

Substituting equation (A.1) into (A.2)

$$\Delta_n = -\frac{\delta_{n-1}}{\sin \alpha_{n-1}} + \delta_n (\cot \alpha_{n-1} + \cot \alpha_n) - \frac{\delta_{n+1}}{\sin \alpha_n}$$

Appendix 8.2

If the length of a column under a compressive load is decreased, the stress at failure will increase and the Euler equation may be used to predict the relationship between the failure stress and the length of column; the modulus of elasticity of the material utilized in the computation is the tangent modulus. If, however, the cross section is made up of a number of thin plate elements, the failure stress will be modified by their localized plate buckling. When the column slenderness ratio is large, the stress for general column failure is less than that at which plate buckling takes place and the Euler equation is used. As the slenderness ratio is reduced, a condition will eventually be reached when both the stress for general column failure and the stress at which the plate element becomes unstable, are equal. Therefore, for slenderness ratios greater than this critical value, general column failure will result and for ratios less than this value, failure is due almost entirely to plate buckling. The stress at this critical slenderness ratio is known as the crippling, crushing or maximum average stress.

References

8.1.   Report of Task Committee on folded plate construction, 'Phase 1 Report on folded plate construction', *J. of the Str. Div. A.S.C.E.,* Dec, 1963.
8.2.   GAAFAR, I., 'Hipped plate analysis considering joint displacement', *Trans. A.S.C.E.,* **119,** 1954.
8.3.   BENJAMIN, B.S. & MAKOWSKI, Z.S., 'The analysis of folded plate structures in plastics', in *Plastics in Building Structures,* Pergamon Press, 1965, pp. 149-163.
8.4.   TIMOSHENKO, S. & WOINOWSKY-KRIEGER, S., *Theory of Plates and Shells.,* McGraw-Hill, 1959.
8.5.   BENJAMIN, B.S., *Structural Design with Plastics,* Van Nostrand Reinhold, 1969.
8.6.·   HOLLAWAY, L. & PARTINGTON, R., 'The analysis and design of a G.R.P. space structure prototype classroom', 2nd Int. Conf. on Space Structures, University of Surrey, Guildford, Surrey, England, Paper 56, 1975.
8.7.   LIVESLEY, R., *Matrix Methods of Structural Analysis,* Pergamon Press, 1964.
8.8.   GILKIE, R.V., 'Pyramids in light weight roof systems', Ph.D. Thesis, University of London, 1967.
8.9.   TIMOSHENKO, S., *Strength of Materials, Part II, Advanced Theory and Problems,* Van Nostrand, 1968.
8.10. SHANLEY, F.R., 'Inelastic column theory', *J. Aero. Sci.,* **14,** 1947
8.11. TIMOSHENKO, S. & GERE, J.M., *Theory of Elastic Stability,* McGraw-Hill, 1961.
8.12. BRYAN, G.H., 'On the stability of a plane plate under thrusts in its own plane and applications to the 'buckling' of the side of a ship', *Proc. Lond. Math. Soc.,* **22,** 1891.
8.13. GERARD, G. & BECKER, H., *Handbook of Structural Stability,* six parts, N.A.C.A. Tech. Notes, pp. 3781–3786, July-August 1957, July 1958.

8.14. BLEICH, F., *Buckling Strength of Metal Structures,* Eng. Soc. Monograph, McGraw-Hill, 1952.

8.15. KORIN, U. & HOLLAWAY, L., 'Compressive characteristics of some pultruded glass reinforced plastics sections', to be published.

8.16. VON KARMAN, T., SECHLER, E.E. & DONNELL, L.H., 'The strength of thin plates in compression', *Trans. Am. Soc. Mech. E.,* June, 1932.

8.17. NEEDHAM, R.A. 'The ultimate strength of aluminium alloy formed structural shapes in compression', *J. Aero. Sci.,* **21**, 4, April, 1954.

8.18. FROST, R.J., 'The ultimate strength of corrugated plates under normal and axial loads', M.Phil. Thesis, University of London, 1968.

8.19. ZIENKIEWICZ, O.C., *The Finite Element Method in Engineering Science,* McGraw-Hill, 1971.

9 Numerical design and detailing examples of G.R.P. components

9.1 Introduction

This chapter is devoted to the practical solution of problems associated with G.R.P. structures and components. The first worked example provides the solution for the longitudinal stresses and transverse bending moments in a prismatic folded plate structure which has been analyzed without the aid of a computer; it will be seen that the analysis is long and tedious. Solutions of problems by the skeletal analysis (section 8.4) and the finite element analysis (section 8.6) techniques have not been included, as both merely require data to be incorporated into the relevant computer program. The results derived from these two methods will be in terms of bending stresses and/or tensile and compressive stresses for the skeletal approach or in terms of inplane and bending stresses within the continuum for the finite element method. Members to resist these forces and moments from either computer analysis would then have to be designed; examples 9.2 to 9.12 illustrate the design procedures.

The values of the ultimate stresses and the modulus of elasticity of the material of the various composites used in the designs have been taken at the lower end of their range. No factors of safety have been incorporated into the design because, as has been pointed out in section 8.8, these are dependent upon the use and life span of the component used and designers must choose the value of these factors most suited to their requirements.

At the end of this chapter details of three G.R.P. structures which have been erected in the U.K. are included.

Example 9.1

This example should be read in conjunction with section 8.2; Figure 9.1.1 should be referred to.

Assume the structure to be manufactured from sandwich construction of dimensions:

  face thickness t = 5 mm
  core thickness c = 25 mm

Edge beam of sandwich construction of dimensions:

  face thickness t = 5 mm
  core thickness c = 50 mm

Determine the longitudinal stresses and transverse bending moments at mid span.

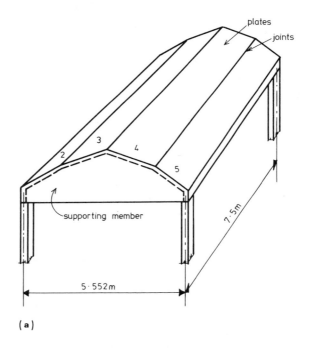

(a)

Figure 9.1.1 Prismatic folded plate structure (a) folded plate structure

Loading

Dead load of folded roof construction sandwich panel and roof covering	=	300 N/m^2
Live load for roof with a slope up to 10°	=	1500 N/m^2
Live load for roof with a slope 30° or less	=	750 N/m^2
Total dead load of edge beam	=	135 N/m^2

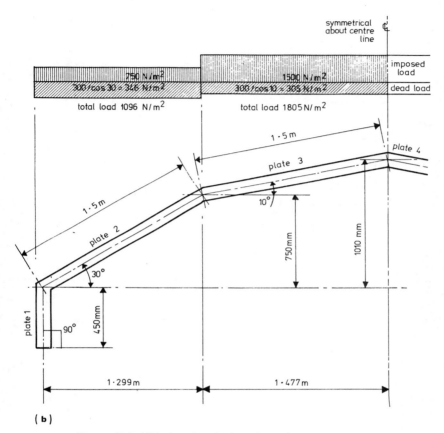

Figure 9.1.1(b) Sectional elevation of structure

Elementary analysis

(a) Transverse slab analysis
The externally applied loads are considered to be carried transversely by plate action.

Table 9.1.1 Properties of the Section

Plate No.	h (mm)	t (mm)	A (mm²)	Z (mm³)	ϕ	$\sin \phi$	$\cos \phi$
1	450	10	4,500	337,500	90°	1.00	0
2	1500	10	15,000	3,750,000	30°	0.50	0.866
3	1500	10	15,000	3,750,000	10°	0.174	0.985

Table 9.1.2

Joint	α	$\sin \alpha$	$\cot \alpha$
0	0	0	∞
1	60	0.866	0.577
2	20	0.342	2.75
3	20	0.342	2.75

Table 9.1.3 Moment Distribution Constants

Joints	Plate	Relative Stiffness	Distribution
0	1		
1	1	$K_{10} = 0$	0
	2	$K_{12} = 4$	1
2	2	$K_{21} = {}^3/_4 \times 4 = 3$	0.429
	3	$K_{23} = 4$	0.571
3	3	$K_{32} = 4$	0.5
	4	$K_{34} = 4$	0.5

Table 9.1.4

Joint	0	1	2		3	
Member		12	21	23	32	34
Distribution factor		1.0	0.429	0.571	0.5	0.5
Fixed end moments (N-m/m)			+231.2	−328.1	+328.1	−328.1
Distribution			+41.6	+55.3	+27.7	
Final moments (N-m/m)			+272.8	−272.8	+355.8	
Moments/span		−210	+210	−56	+56	
$\dfrac{w \times \text{span}}{2}$ (N)	+135	+712	+712	+1333	+1333	
Total shear (N)	+135	+502	+922	+1277	+1389	
Joint reactions (N)	637		2199		2778	

A typical 1 m strip of slab continuous over the supports is analyzed by moment distribution. The fixed end moments are $WL^2/12$

The fixed end moment $M_{21} = 1.5 \times \dfrac{1096 \times 1.299^2}{12}$

$$= 231.2 \text{ N}-\text{m}$$

The fixed end moment $M_{23} = M_{32} = \dfrac{1805 \times 1.477^2}{12}$

$$= 328.1 \text{ N}-\text{m}$$

(b) Longitudinal plate analysis

Plate loads—the vertical joint reactions are resolved into components in the plane of each plate.

Plate 1

$$P_0 = 0$$

$$P_1 = 637$$

Plate 2

$$P_1 = 0$$

$$P_2 = \frac{2199}{\sin 20} \sin 100 = +6332 \text{ N/m}$$

Plate 3

$$P_2 = -\frac{2199}{\sin 20} \sin 60 = -5568 \text{ N/m}$$

$$P_3 = \frac{2778}{\sin 20} \sin 80 = +7999 \text{ N/m}$$

Free edge stresses. The longitudinal stresses developed by these plates, which are assumed to deflect independently, are derived from plate loads and are shown plotted in Figure 9.1.2a.

load in plate 3 = $- 5568 + 7999 = 2431$ N/m

moment at centre of plate
$$= \frac{WL^2}{8} = + \frac{2431 \times 7.5^2}{8} = + 17 \ 093 \text{ N}-\text{m/m}$$

$$\sigma_t = -\sigma_b = \frac{17093 \times 10^3}{3750000} = 4.56 \text{ N/mm}^2$$

load in plate 2 $= + 6332$ N/m

moment at centre of plate
$$= \frac{WL^2}{8} = \frac{6332 \times 7.5^2}{8} = 44 \ 522 \text{ N}-\text{m/m}$$

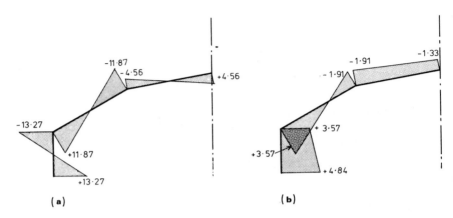

Figure 9.1.2 Longitudinal stresses (N/mm²) at mid-span from the elementary analysis; (a) free edge stresses; (b) distributed free edge stresses

$$\sigma_t = -\sigma_b = -\frac{44\,522 \times 10^3}{3750000} = -11.87 \text{ N/mm}^2$$

load in plate 1 $= 637.0$ N/m

moment at centre of plate

$$= \frac{637 \times 7.5^2}{8} = 4479 \text{ N-m}$$

$$\sigma_t = -\sigma_b = \frac{4479 \times 10^3}{337500} = -13.27 \text{ N/mm}^2$$

Free edge stress distribution. The edge stresses must be equal at each joint. The equalized edge stresses will be obtained by distributing the unbalanced free edge stresses at each joint similarly to that of unbalanced fixed end moments in moment distribution. The free edge stresses are mathematically analogous to fixed end moments; the relative reciprocals of plate areas are analogous to the relative stiffness factors and the carry over factor of $-\frac{1}{2}$ is analogous to $+\frac{1}{2}$ in moment distribution. The distribution is shown in Table 9.1.5, and the free edge stresses are plotted in Figure 9.1.2b.

Stress distribution constants

$$\text{Joint 1} \quad D_{10} = \frac{A_1}{A_1 + A_2} = \frac{4500}{4500 + 15000} = 0.231$$

$$D_{12} = \frac{A_2}{A_1 + A_2} = \frac{15000}{4500 + 15000} = 0.769$$

$$\text{Joint 2} \quad D_{21} \; = \; \frac{A_2}{A_2 + A_3} \; = \; \frac{15000}{15000 + 15000} = 0.5$$

$$D_{23} \; = \; \frac{A_3}{A_2 + A_3} \; = \; \frac{15000}{15000 + 15000} = 0.5$$

Table 9.1.5

Joint	0		1		2		3
Member	01	10	12	21	23		32
Distribution factor		0.769	0.231	0.5	0.5		0.5
Free edge stress (N/mm²)	+13.27	−13.27	+11.87	−11.87	+4.56		−4.56
Distribution		+19.33	−5.81	+8.22	−8.22		
Carry-over	−9.67		−4.11	+2.90			+4.11
Distribution		−3.16	+0.95	−1.45	+1.45		
Carry-over	+1.58		+0.73	+0.48			−0.73
Distribution		+0.56	−0.17	−0.24	+0.24		
Carry-over	−0.28		+0.12	+0.08			−0.12
Distribution		+0.09	−0.03	−0.04	+0.04		
Carry-over	−0.05		+0.02	+0.02			−0.02
Distribution		+0.02	0	−0.01	+0.01		
(N/mm²)	+4.84	+3.57	+3.57	−1.91	−1.91		1.32

Plate deflection

The direction for positive plate deflection σ is given in Figure 9.1.3.

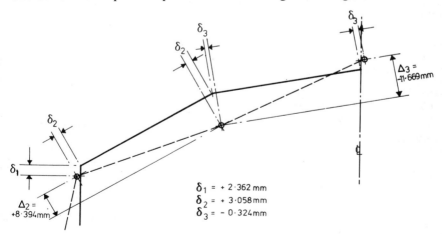

$$\delta_1 = + 2 \cdot 362 \text{ mm}$$
$$\delta_2 = + 3 \cdot 058 \text{ mm}$$
$$\delta_3 = - 0 \cdot 324 \text{ mm}$$

$$\Delta_2 = +8 \cdot 394 \text{ mm}$$
$$\Delta_3 = -11 \cdot 669 \text{ mm}$$

Figure 9.1.3 Geometric relationship between plate deflections and relative joint displacements

The modulus of elasticity of the faces is 7.0 GN/m². The centre span deflection of a simply supported beam under uniform loading is:

$$\delta = \frac{5\, ML^2}{48\, EI}$$

where $M = \dfrac{\sigma_b - \sigma_t}{2}\, Z$

so $\delta = \dfrac{5(\sigma_b - \sigma_t)L^2\, Z}{96\, EI}$

where $\dfrac{Z}{I} = \dfrac{2}{h}$ for rectangular beam

$$\delta = \frac{(\sigma_b - \sigma_t)}{h}\ \frac{5}{48}\ \frac{L^2}{E}$$

$$\delta_{3e} = \frac{-1.91 + 1.33}{1.5 \times 10^3} \times \frac{5}{48} \times \frac{7.5^2 \times 10^6 \times 10^6}{7 \times 10^9}$$

$$= -0.324\ \text{mm}$$

$$\delta_{2e} = \frac{(3.57 + 1.91)}{1.5 \times 10^3} \times \frac{5}{48} \times \frac{7.5^2 \times 10^6 \times 10^6}{7 \times 10^9}$$

$$= +3.058\ \text{mm}$$

$$\delta_{1e} = \frac{(4.84 - 3.57)}{450} \times \frac{5}{48} \times \frac{7.5^2 \times 10^6 \times 10^6}{7 \times 10^9}$$

$$= +2.362\ \text{mm}$$

Relative joint displacement

The relationship between plate deflections and relative joint displacements is given by:

$$\Delta_n = -\frac{\delta_{n-1}}{\sin \alpha_{n-1}} + \delta_n\, (\cot \alpha_{n-1} + \cot \alpha_n) - \frac{\delta_{n+1}}{\sin \alpha_n}$$

This equation must be evaluated for each plate whose relative joint displacement causes stresses.

The geometrical relationship between these two quantities is given in Figure 9.1.3.

$$\Delta_2 = -\frac{\delta_1}{\sin 60} + \delta_2 (\cot 60 + \cot 20) - \frac{\delta_3}{\sin 20}$$

$$= -1.155 \, \delta_1 + 3.324 \, \delta_2 - 2.924 \, \delta_3$$

$$\Delta_3 = -\frac{\delta_2}{\sin 20} + \delta_3 (\cot 20 + \cot 20) - \frac{\delta_4}{\sin 20}$$

$$= -2.924 \, \delta_2 + 5.494 \, \delta_3 - 2.924 \, \delta_4$$

$$\Delta_{2e} = -1.155 \times 2.362 + 3.324 \times 3.058 + 2.924 \times 0.324$$

$$= 8.394 \text{ mm}$$

$$\Delta_{3e} = -2.924 \times 3.058 - 5.494 \times 0.324 - 2.924 \times 0.324$$

$$= -11.669 \text{ mm}$$

These displacements indicate a transverse distortion of the cross section which is contrary to the assumption, inherent in the elementary analysis, that all joints deflect equally.

Correction analysis

The correction analysis is designed to deal with the transverse distortion of the cross section arising from the previous analysis. Since all plastics are assumed to have no torsional resistance plates 1 and 6 will rotate freely in the transverse direction and plates 2, 3, 4 and 5 are restrained by each other and therefore cannot freely rotate. By symmetry plates 2 and 3 will rotate the same amount as 5 and 4 respectively; consequently, only two plates will be considered.

(a) *Transverse slab analysis*

Figure 9.1.4 shows the arbitrary relative joint displacements imposed and the resulting fixed end moments.

Considering a displacement of 1 mm at joint 1

Moment at joint $2 = 3 \, E_f \, I \, \Delta/h^2$

$$= \frac{3 \times 7 \times 10^9 \times 2250 \times 1}{1.5 \times 10^3 \times 1.5 \times 10^3 \times 10^6}$$

$$= 21.0 \text{ N-mm/mm}$$

where $\quad I = \dfrac{5 \times 1 \times 30^2}{2}$

$$= 2250 \text{ mm}^4$$

$$E_f = 7.0 \text{ GN/m}^2$$

Considering a displacement of 1mm at joint 2

Moment at joint 3 $= 6\,E_f\,I\,\Delta/h^2$

$= 42.0$ N$-$mm/mm

The transverse moments, shears and joint reactions may be computed by a moment distribution for:

$\Delta_2 = 1$ mm.

Table 9.1.6

Joints	0	1		2			3
Members		12	21	23		32	
Distribution factors		1.0	0.429	0.571		0.5	0.5
Fixed end moment (N-mm/mm) Distribution Carry-over			+9.009	−21.0 +11.991		+5.996	
Final moment (N-mm/mm)			−11.991	+11.991		+5.996	
Moment/span		+9.23	−9.23	−12.178		+12.178	
Joint reaction (N × 10⁻³/mm)		+9.23 (upwards)		−21.408 (downward)		+24.356 (upward)	

$\Delta_3 = 1$ mm

Table 9.1.7

Joints	0	1		2			3
Members		12	21	23		32	
Distribution factors		1.0	0.429	0.571			
Fixed end moments (N-mm/mm) Distribution			+18.018	−42.0 +23.982		−42.0	
Carry-over						+11.991	
Final moment (N-mm/mm) Moment/span (N × 10⁻³/mm)		−13.871	+18.018 +13.871	−18.018 +32.517		−30.009 −32.517	
Joint reaction (N × 10⁻³/mm)		−13.871		+46.388		−65.034	

These joint reactions are converted to plate loads and then to free edge stresses in a similar manner to the elementary analysis.

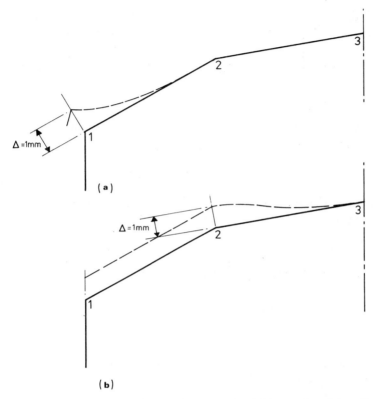

Figure 9.1.4 Arbitrary relative joint displacements imposed at joints 2 and 3; (a) bending moment at joint $2 = 21.0$ N-mm/mm; (b) bending moment at joint $3 = 42.0$ N-mm/mm

(b) *Longitudinal plate analysis*

For $\quad \Delta_2 = 1$ mm

Plate 1

$\quad P_0 = 0$

$\quad P_1 = 9.23 \times 10^{-3}$ N/mm

Plate 2

$\quad P_1 = 0$

$\quad P_2 = -21.408 \times 10^{-3} \times \dfrac{\sin 100}{\sin 20} = 61.642 \times 10^{-3}$ N/mm

Plate 3

$$P_2 = +21.408 \times 10^{-3} \times \frac{\sin 60}{\sin 20} = 54.207 \times 10^{-3} \text{ N/mm}$$

$$P_3 = 24.356 \times 10^{-3} \times \frac{\sin 80}{\sin 20} = 70.130 \times 10^{-3} \text{ N/mm}$$

So load in Plate 3

$$= (70.130 + 54.207) \times 10^{-3} = 124.337 \times 10^{-3} \text{ N/mm}$$

load in Plate 2 $\qquad = -61.642 \times 10^{-3} \text{ N/mm}$

load in Plate 1 $\qquad = 9.23 \times 10^{-3} \text{ N/mm}$

The free edge stresses are obtained from equation (8.3) and have the following values:

For Plate 3

$$\sigma_b = -\sigma_t = \frac{124.337 \times 10^{-3} \times 7500^2}{3750000 \times \pi^2} \text{ N/mm}^2$$

$$= +0.189 \text{ N/mm}^2$$

For Plate 2

$$\sigma_b = -\sigma_t = \frac{-61.642 \times 10^{-3} \times 7500^2}{3750000 \times \pi^2}$$

$$= -0.0937 \text{ N/mm}^2$$

For Plate 1

$$\sigma_b = -\sigma_t = \frac{9.23 \times 10^{-3} \times 7500^2}{337500 \times \pi^2}$$

$$= 0.1559 \text{ N/mm}^2$$

For $\Delta_3 = 1$ mm

Plate 1

$$P_0 = 0$$

$$P_1 = -13.871 \times 10^{-3} \text{ N/mm}$$

Plate 2

$$P_1 = 0$$

Numerical Design and Detailing Examples of G.R.P. Components

$$P_2 = +46.388 \times 10^{-3} \times \frac{\sin 100}{\sin 20} = +133.569 \times 10^{-3} \text{ N/mm}$$

Plate 3

$$P_2 = -46.388 \times 10^{-3} \times \frac{\sin 60}{\sin 20} = -117.459 \times 10^{-3} \text{ N/mm}$$

$$P_3 = -65.034 \times 10^{-3} \times \frac{\sin 80}{\sin 20} = -187.258 \times 10^{-3} \text{ N/mm}$$

So load in Plate 3

$$= (-117.459 - 187.258) \times 10^{-3} = -304.717 \times 10^{-3} \text{ N/mm}$$

load in Plate 2 $= +133.569 \times 10^{-3} \text{ N/mm}$

load in Plate 1 $= -13.871 \times 10^{-3} \text{ N/mm}$

The free edge stresses are:

For Plate 3

$$\sigma_b = -\sigma_t = \frac{-304.717 \times 10^{-3} \times 7500^2}{3750000 \times \pi^2}$$

$$= -0.4631 \text{ N/mm}^2$$

For Plate 2

$$\sigma_b = -\sigma_t = \frac{+133.569 \times 10^{-3} \times 7500^2}{3750000 \times \pi^2}$$

$$= +0.2030 \text{ N/mm}^2$$

For Plate 1

$$\sigma_b = -\sigma_t = \frac{-13.871 \times 10^{-3} \times 7500^2}{337500 \times \pi^2}$$

$$= -0.2342 \text{ N/mm}^2$$

These free edge stresses show incompatibility which may be removed by stress distribution.

The plate deflections are computed from equation (8.5)

$$\delta = \frac{\sigma_b - \sigma_t}{E h} \frac{L^2}{\pi^2}$$

for $\Delta_2 = 1$ mm

$$\delta_3 = \frac{(14.37 + 16.63) \times 10^{-2} \times 7500^2}{7 \times 10^3 \times 1500 \times \pi^2}$$

$$= 0.1683 \text{ mm}$$

N

Table 9.1.8 Distribution of stresses resulting from arbitrary relative joint displacements of 1 mm at mid-span

$\Delta_2 = 1$ mm.

Joint	0		1		2	3
Distribution factor	0	0.769	0.231	0.5	0.5	0.5
Carry-over factor		$-^1/_2$		$-^1/_2$		$-^1/_2$
Free edge stress	+15.59	−15.59	−9.37	+9.37	+18.90	−18.90
(N/cm²)		+4.78	−1.44	+4.77	−4.76	0
Carry-over	−2.39		−2.39	+0.72		+2.38
Distribution		−1.84	+0.55	−0.36	+0.36	0
Carry-over	+0.92		+0.18	−0.28		−0.18
Distribution		+0.14	−0.04	+0.14	−0.14	0
Carry-over	−0.07		−0.07	+0.02		+0.07
Distribution	+0.02	−0.05	+0.02	−0.01	+0.01	
Total Stress	+14.07	−12.56	−12.56	+14.37	+14.37	−16.63
(N/cm²)						

$\Delta_3 = 1$ mm.

Table 9.1.9

Joint	0		1		2	3
Distribution factor	0	0.769	0.231	0.5	0.5	
Carry-over		$-\frac{1}{2}$		$-\frac{1}{2}$		$-\frac{1}{2}$
Free edge stress	−23.42	+23.42	+20.30	−20.30	−46.31	+46.31
N/cm²						
Distribution	0	−2.40	+0.72	−13.01	+13.00	0
Carry-over	+1.20		+6.51	−0.36		−6.50
Distribution		+5.01	−1.50	+0.18	−0.18	0
Carry-over	−2.51	0	−0.09	+0.75	0	+0.09
Distribution	0	−0.07	+0.02	−0.38	+0.37	
Carry-over	+0.03		+0.19			−0.18
Distribution		+0.15	−0.04			
Total N/cm² Stress	−24.70	+26.11	+26.11	−33.12	−33.12	+39.72

$$\delta_2 = \frac{(-12.56 - 14.37) \times 10^{-2} \times 7500^2}{7 \times 10^3 \times 1500 \times \pi^2}$$

$$= -0.1462 \text{ mm}$$

$$\delta_1 = \frac{(14.07 + 12.56) \times 10^{-2} \times 7500^2}{7 \times 10^3 \times 450 \times \pi^2}$$

$$= 0.4818 \text{ mm}$$

for $\Delta_3 = 1$ mm

$$\delta_3 = \frac{(-33.12 - 39.72) \times 10^{-2} \times 7500^2}{7 \times 10^3 \times 1500 \times \pi^2}$$

$$= -0.3954 \text{ mm}$$

$$\delta_2 = \frac{(+26.11 + 33.12) \times 10^{-2} \times 7500^2}{7 \times 10^3 \times 1500 \times \pi^2}$$

$$= +0.3215 \text{ mm}$$

$$\delta_1 = \frac{(-24.70 - 26.11) \times 10^{-2} \times 7500^2}{7 \times 10^3 \times 450 \times \pi^2}$$

$$= -0.9193 \text{ mm}$$

The total centre deflection of the plates is equal to the summation of the deflection obtained from the elementary analysis $\delta_{1e}, \delta_{2e}, \delta_{3e}$ and the product of the preceding deflections obtained from equation (8.5) and the actual relative joint displacements Δ_2 and Δ_3

$$\delta_1 = +2.362 + 0.4818 \Delta_2 - 0.9193 \Delta_3$$

$$\delta_2 = +3.058 - 0.1462 \Delta_2 + 0.3215 \Delta_3$$

$$\delta_3 = -0.324 + 0.1683 \Delta_2 - 0.3954 \Delta_3$$

$$\delta_4 = +0.324 - 0.1683 \Delta_2 + 0.3954 \Delta_3$$

These expressions for plate deflections are substituted into equation (8.1) for geometric compatibility

$$\Delta_2 = -1.155 \delta_1 + 3.324 \delta_2 - 2.924 \delta_3$$

$$\Delta_3 = -2.924 \delta_2 + 5.494 \delta_3 - 2.924 \delta_4$$

from which

$$\Delta_2 = +0.7982 \text{ mm}$$

$$\Delta_3 = -1.9355 \text{ mm}$$

The elementary and corrected analyses are combined in Table 9.1.10 to give the final values of the horizontal stresses at mid-span and in Table 9.1.11 to give the final values of the transverse bending moments at mid-span.

Table 9.1.10 Stresses in N/mm^2 + ve and – ve indicate tension and compression respectively
Longitudinal stresses at mid-span

Joints	0		1		2		3
Elementary *analysis* (N-m/m)	+4.84		+3.57	+3.57	−1.91	−1.91	−1.33
Corrected analysis							
$\Delta_2 = 1$ mm	+0.1407		−0.1256	−0.1256	+0.1437	+0.1437	−0.1663
$\Delta_2 = 0.7982$	+0.1123		−0.1003	−0.1003	+0.1147	+0.1147	−0.1327
$\Delta_3 = 1$ mm	−0.2470		+0.2611	+0.2611	−0.3312	−0.3312	+0.3972
$\Delta_3 = -1.9355$	+0.4781		−0.5054	−0.5054	+0.6410	+0.6410	−0.7688
Total Correction (N-m/m)	+0.5904		−0.6057	−0.6057	+0.7557	+0.7557	−0.9015
Final values (N-m/m)	+5.4304		+2.9643	+2.9643	−1.1543	−1.1543	−2.2315

Table 9.1.11 Transverse bending moments at mid-span

	Joint 1	Plate 2	Joint 2	Plate 3	Joint 3
Elementary *analysis* (N-m/m)	0	+94.8	−272.8	+177.85	−355.8
Corrected Analysis					
$\Lambda_2 = 1$ mm	0	+5.996	+11.991	+2.998	−5.996
$\Delta_2 = 0.7982$		+4.786	+9.571	+2.393	−4.786
$\Delta_3 = 1$ mm	0	−9.009	−18.018	+5.996	+30.009
$\Delta_3 = -1.9355$		+17.439	+34.874	−11.605	−58.082
Total correction (N-m/m)	0	+22.223	+44.445	−9.212	−62.868
Final values (N-m/m)	0	117.023	−228.355	+168.638	−418.668

It is now necessary to derive general expressions for the longitudinal shearing forces so that the shearing stresses at any point in the folded plate system may be computed. By combining the longitudinal stresses (Table 9.1.10) with the computed longitudinal shear stresses, the longitudinal principal stresses are obtained. These calculations are left to the reader.

The significant results of the analysis are the transverse bending moments and the longitudinal principal stresses from which the transverse and longitudinal composite plates' dimensions respectively may be determined.

The following examples are based upon the theory discussed in Part II of Chapter 8 and they illustrate the design of component members of structural configurations.

Example 9.2

This example utilizes the law of mixtures as given in section 2.4.1.1.

It is required to obtain the stress at the point of matrix rupture and elastic modulus of a composite consisting of unidirectional glass fibres in a polyester matrix under a tensile force (see Figure 9.2.1). The composite has the following geometric and mechanical properties.

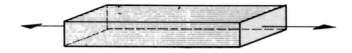

Figure 9.2.1

Resin ratio by volume $= 65\%$
Area of composite $= 25 \text{ mm} \times 2 \text{ mm} = 50 \text{ mm}^2$
The ultimate stress in the matrix $= 62.0 \text{ MN/m}^2$
The modulus of elasticity of the glass fibres $= 70 \text{ GN/m}^2$
The modular ratio of the composite $= 20$

Area of glass in composite $= 50 \times 0.35 = 17.5 \text{ mm}^2$
Area of matrix $= 50 \times 0.65 = 32.5 \text{ mm}^2$
Equation (2.2) states
$$\sigma_c A_c = \sigma_m A_m + \sigma_f A_f$$

$$50\sigma_c = \sigma_m 32.5 + \sigma_f 17.5 \tag{9.2.1}$$

Equation (2.4) states
$$\epsilon_c = \epsilon_m = \epsilon_f$$

$$\frac{\sigma_c}{E_c} = \frac{\sigma_m}{E_m} = \frac{\sigma_f}{E_f}$$

$$\therefore \sigma_f = \sigma_m \frac{E_f}{E_m}$$

$$= \sigma_m \, 20 \tag{9.2.2}$$

Substituting equation (9.2.2) into equation (9.2.1)

$$\sigma_c = \frac{62.0}{50} [32.5 + 17.5 \times 20]$$

$$= 474.3 \text{ MN/m}^2$$

The composite elastic modulus

$$E_c = E_m V_m + E_f V_f$$

$$= \frac{70}{20} \times 0.65 + 70 \times 0.35$$

$$= 26.775 \text{ GN/m}^2$$

Example 9.3

It is required to determine the ultimate load carrying capacity of a composite material (defined here as one consisting of two or more dissimilar materials which themselves are composites) per metre width under a tensile load P. The composite material is composed of:

One lamina (defined as a material consisting of both fibres and matrix) (composite) of thickness 3 mm consisting of glass fibre woven roving reinforcement and polyester resin.

One lamina (composite) of thickness 6 mm consisting of chopped strand mat reinforcement and polyester resin.

One lamina (composite) of thickness 1.5 mm consisting of glass cloth reinforcement and polyester resin.

The composite is shown in Figure 9.3.1.

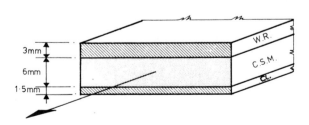

Figure 9.3.1

The ultimate values of tensile stress and modulus of elasticity for the laminae are given in Table 9.3.1.

The line of application of the tensile load is parallel to the direction of the main reinforcement in the woven roving and cloth laminates.

Table 9.3.1 Values of tensile stress and modulus of elasticity for glass fibre polyester resin composite

		Mat	Cloth	Woven roving
Ultimate tensile stress	MN/m²	65.0	123.0	164.0
Modulus of elasticity	GN/m²	6.2	9.7	9.2

Strains in all layers at a particular cross section are equal

$$\therefore \quad \epsilon_{csm} = \epsilon_{cl} = \epsilon_{wr}$$

$$\frac{\sigma_{csm}}{E_{csm}} = \frac{\sigma_{cl}}{E_{cl}} = \frac{\sigma_{wr}}{E_{wr}} \tag{9.3.1}$$

$$\sigma_{csm} = E_{csm} \frac{\sigma_{cl}}{E_{cl}} = E_{csm} \frac{\sigma_{wr}}{E_{wr}} \tag{9.3.2}$$

$$\sigma_{cl} = E_{cl} \frac{\sigma_{csm}}{E_{csm}} = E_{cl} \frac{\sigma_{wr}}{E_{wr}} \tag{9.3.3}$$

$$\sigma_{wr} = E_{wr} \frac{\sigma_{csm}}{E_{csm}} = E_{wr} \frac{\sigma_{cl}}{E_{cl}} \tag{9.3.4}$$

From equations (9.3.2), (9.3.3) and (9.3.4) when the

(a) ultimate value of σ_{csm} = 65.0 MN/m²

σ_{cl} = 101.69 MN/m²

σ_{wr} = 96.45 MN/m²

(b) ultimate value of σ_{cl} = 123.0 MN/m²

σ_{csm} = 78.62 MN/m²

σ_{wr} = 116.66 MN/m²

(c) ultimate value of σ_{wr} = 164.0 MN/m²

σ_{csm} = 110.52 MN/m²

σ_{cl} = 172.91 MN/m²

Part (a) states that when the ultimate stress for the C.S.M. laminate is reached the stress values in the cloth and woven roving laminates are less than their ultimate.

Part (b) states that when the ultimate stress for the cloth laminate is reached the stress value in the woven roving laminate is less than its ultimate but the stress value in the C.S.M. laminate has been exceeded.

Part (c) states that when the ultimate stress for the woven roving laminate is reached the ultimate stress values in the C.S.M. and cloth laminates have been exceeded.

Therefore the criteria of failure of the composite material is the C.S.M. laminate and the ultimate load carrying capacity of the composite is:

$$(65.0 \times 6 \times 10^3 + 101.69 \times 1.5 \times 10^3 + 96.45 \times 3 \times 10^3) \text{ N}$$
$$= 831.885 \text{ kN}$$

A check should be made of the ultimate load on the composite material after the failure of the C.S.M. laminate. In this case the ultimate load the composite material is able to support is:

$$(123 \times 1.5 \times 10^3 + 116.66 \times 3 \times 10^3) \text{ N}$$
$$= 534.48 \text{ kN}$$

Example 9.4

A composite consists of woven roving and polyester resin. Its ultimate tensile strength in the warp and weft directions and at 45° to these are:

$$\sigma_1 = \sigma_{11} = 288.0 \text{ MN/m}^2$$

$$\sigma_t = \sigma_{22} = 260.0 \text{ MN/m}^2$$

$$\sigma_{45} = \sigma_{xx} = 81.4 \text{ MN/m}^2$$

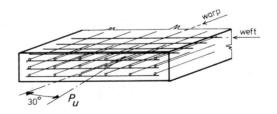

Figure 9.4.1

It is required to determine the ultimate tensile strength of the composite at 30° to the warp direction (see Figure 9.4.1).

Considering equation (8.22)

$$\frac{1}{\sigma_s^2} = \frac{1}{\sin^2\theta \cos^2\theta} \left[\frac{1}{\sigma_{xx}^2} - \frac{\cos^4\theta}{\sigma_{11}^2} - \frac{\sin^4\theta}{\sigma_{22}^2} \right]$$

$$\sigma_s = 41.64 \text{ MN/m}^2 \quad (\text{at } \theta = 45^0)$$

Therefore stress at 30^0

$$\frac{1}{(\sigma_{30})^2} = \frac{\cos^4\theta}{\sigma_{11}^2} + \frac{\sin^4\theta}{\sigma_{22}^2} + \frac{\sin^2\theta\,\cos^2\theta}{\sigma_s^2}$$

$\sigma_{30} = 92.91$ MN/m² (at $\theta = 30^0$)

Example 9.5

A composite material is manufactured from a lamina (composite) of chopped strand mat reinforcement and polyester resin of thickness 10 mm sandwiched between two glass cloth reinforcement and polyester resin laminates (composites) of 1.5 mm thickness. It is required to find the total ultimate tensile load/metre width that the composite can carry when the line of action of the applied tensile load makes an angle of 15° to the warp direction of the cloth laminates (see Figure 9.5.1).

The ultimate values of the tensile stress and modulus of elasticity are given in Table 9.5.1.

Table 9.5.1

Ultimate tensile values (MN/m²)	Chopped strand mat	Cloth
σ_{11}	65.0	123.0
σ_{22}	65.0	95.6
$\sigma_{xx} = \sigma_{45}$	65.0	58.0
Modulus of elasticity (GN/m²)		
E_{11}	7.0	13.45
E_{22}	7.0	12.41
G_{12}		3.59
v_{12}		0.20

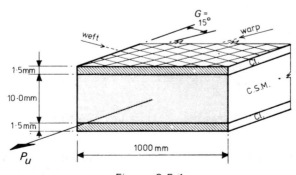

Figure 9.5.1

Chopped strand mat laminate is assumed to be isotropic in behaviour. Cloth laminate is assumed to behave as an orthotropic material, therefore the values of the ultimate stress and modulus of elasticity at 15° to the warp direction must be calculated.

From equation (8.22)

$$\frac{1}{\sigma_s^2} = \frac{1}{\sin^2\theta \cos^2\theta} \left[\frac{1}{\sigma_{xx}^2} - \frac{\cos^4\theta}{\sigma_{11}^2} - \frac{\sin^4\theta}{\sigma_{22}^2} \right]$$

σ_s = 31.41 MN/m² (at $\theta = 45°$)

Also from equation (8.22)

$$\frac{1}{\sigma_{xx}^2} = \frac{\cos^4\theta}{\sigma_{11}^2} + \frac{\sin^4\theta}{\sigma_{22}^2} + \frac{\sin^2\theta \cos^2\theta}{\sigma_s^2}$$

where σ_{xx} = stress at 15° to the warp direction

$\theta = 15°$ to the warp direction

from which σ_{xx} = 90.77 MN/m²

From equation (2.27)

$$\frac{1}{E_{xx}} = \frac{\cos^4\theta}{E_{11}} + \frac{\sin^4\theta}{E_{22}} + \left[\frac{1}{G_{12}} - \frac{2\nu_{12}}{E_{11}} \right] \cos^2\theta \sin^2\theta$$

where E_{xx} = modulus of elasticity at 15° to the warp direction

$\theta = 15°$ to the warp direction

from which E_{xx} = 12.40 GN/m²

Strain in all layers of composite material at 15° to the warp direction is:

$\epsilon_{cl} = \epsilon_{csm}$

$$\frac{\sigma_{cl}}{E_{cl}} = \frac{\sigma_{csm}}{E_{csm}}$$

so $\sigma_{csm} = E_{csm} \times \dfrac{\sigma_{cl}}{E_{cl}} = 90.77 \times \dfrac{7.0}{12.40} = 51.24$ MN/m²

and $\sigma_{cl} = E_{cl} \times \dfrac{\sigma_{csm}}{E_{csm}} = 65.0 \times \dfrac{12.40}{7.0} = 115.14$ MN/m²

Therefore the ultimate stress in the cloth laminate is the critical value for the composite; the stress in the chopped strand mat is 51.24 MN/m², and the stress in the cloth laminate is 90.77 MN/m².

∴ Total load applied to the composite is

$= A_{csm}\,\sigma_{csm} + A_{cl}\,\sigma_{cl}$

$= (10 \times 51.24 + 3 \times 90.77)\,10^{-3}$ MN/m width

$= (512.4 + 272.31)$ MN/m width

$= 784.71$ MN/m width

Example 9.6

It is required to determine the critical compressive load for a chopped strand mat reinforcement and polyester resin composite of width 150 mm, thickness 12 mm and length 375 mm. The strut is simply supported at its ends. It is assumed that for this composite the reduced modulus E_r was determined by experiment and had the value 6.9 GN/m² (see Figure 9.6.1).

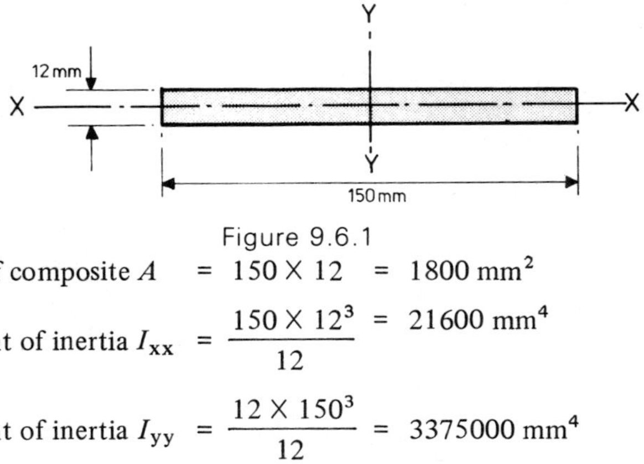

Figure 9.6.1

Area of composite A $= 150 \times 12$ $= 1800$ mm²

Moment of inertia $I_{xx} = \dfrac{150 \times 12^3}{12} = 21600$ mm⁴

Moment of inertia $I_{yy} = \dfrac{12 \times 150^3}{12} = 3375000$ mm⁴

Least radius of gyration $r_{yy} = \sqrt{\left[\dfrac{21600}{1800}\right]} = 3.464$ mm

Slenderness ratio $\dfrac{l}{r_{yy}} = \dfrac{375}{3.464} = 108.26$

∴ the critical buckling load $= \dfrac{\pi^2 EA}{(l/r_{yy})^2}$

$= \dfrac{\pi^2 \times 6.9 \times 1800 \times 10^3}{(108.26)^2} = 10458.88$ N

Example 9.7

A composite material strut of width 150 mm and of length 600 mm is shown in Figure 9.7.1 and is composed of:

One lamina (composite) of thickness 3 mm consisting of glass fibre woven roving reinforcement and polyester resin;

One lamina (composite) of thickness 9 mm consisting of chopped strand mat reinforcement and polyester resin;

One lamina (composite) of thickness 1.5 mm consisting of glass cloth reinforcement and polyester resin.

It is required to determine the critical compressive load parallel to the warp direction for simply supported end conditions.

The reduced modulus of elasticity for woven roving composite = $16.8 \, GN/m^2$;

The reduced modulus of elasticity for chopped strand mat composite = $6.9 \, GN/m^2$;

The reduced modulus of elasticity for cloth composite = $17.0 \, GN/m^2$;

These values have been determined by experiment.

Area of woven roving composite = $150 \times 3 = 450 \, mm^2$
Area of chopped strand mat composite = $150 \times 9 = 1350 \, mm^2$
Area of cloth composite = $150 \times 1.5 = 225 \, mm^2$

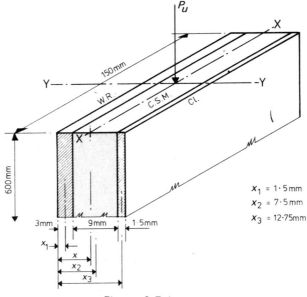

Figure 9.7.1

The distance of the centroid of the composite material from the free edge of the woven roving composite is

$$\bar{x} = \frac{\Sigma A_i E_i x_i}{\Sigma EA}$$

$$= \frac{(17.0 \times 225 \times 12.75) + (6.9 \times 1350 \times 7.5) + (16.8 \times 450 \times 1.5)}{(17.0 \times 225) + (6.9 \times 1350) + (16.8 \times 450)}$$

$$= 6.279 \text{ mm}$$

The composite modulus of elasticity E value is given by:

$$E = \frac{1}{\Sigma A} \Sigma E_i A_i$$

$$= \frac{(16.8 \times 450) + (6.9 \times 1350) + (225 \times 17.0)}{450 + 1350 + 225}$$

$$= 10.22 \text{ GN/m}^2$$

The second moment of area of each laminate about its own centroidal axis

woven roving composite $= \dfrac{150 \times 3^3}{12} = 337.5 \text{ mm}^4$

chopped strand mat composite $= \dfrac{150 \times 9^3}{12} = 9112.5 \text{ mm}^4$

cloth composite $= \dfrac{150 \times 1.5^3}{12} = 42.19 \text{ mm}^4$

The stiffness of the composite material is

$$EI_{xx} = \Sigma E_i (I_i + A_i x_i^2)$$

$$= 16.8 (337.5 + 450 \times 4.779^2) + 6.9 (9112.5 + 1350 \times 1.221^2)$$

$$+ 17.0 (42.19 + 225 \times 6.471^2)$$

$$EI_{xx} = 415{,}979.744 \text{ GN-mm}^4/\text{m}^2$$

$$I_{xx} = 40702.519 \text{ mm}^4$$

Least radius of gyration $r_{yy} = \sqrt{\left[\dfrac{40702.519}{2025.0} \right]} = 4.4833 \text{ mm}$

slenderness ratio $\dfrac{l}{r_{yy}} = \dfrac{600}{4.4833}$

$$= 133.83$$

$$\therefore \text{ critical load} = \frac{\pi^2 EA}{(l/r)^2}$$

$$= \frac{\pi^2 \times 10.22 \times 2025.0}{133.83^2}$$

$$= 11.404 \times 10^3 \text{ N}$$

Example 9.8

It is required to determine the ultimate load carrying capacity per 10 mm width of a simply supported beam (of composite material) when under a uniformly distributed load. The composite material is manufactured from:

One lamina (composite) of thickness 1 mm consisting of glass cloth reinforcement and polyester resin;

One lamina (composite) of thickness 1.5 mm consisting of chopped strand mat reinforcement and polyester resin;

One lamina of thickness 4 mm consisting of woven roving reinforcement and polyester resin.

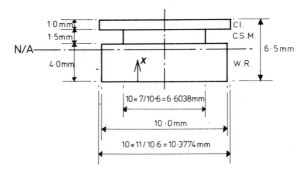

Figure 9.8.1

Table 9.8.1.

Reinforcement	Cloth	Chopped strand mat	Woven roving
The modulus of elasticity in bending (GN/m²)	11.0	7.0	10.6
Ultimate strength in bending (MN/m²)	216.0	140.0	195.0
Ultimate interlaminar shear stress (MN/m²)	19.0	22.0	20.0

The warp direction of the woven roving and cloth reinforcements are parallel to the direction of the span, which is 300 mm (see Figure 9.8.1).

Table 9.8.1 gives the mechanical properties of the three laminae.

As the mechanical properties of the three laminates are different the analysis for a composite section must be used.
The modulus of section is obtained from

$$Z = \frac{EI}{E_i x}$$

where x = distance from the neutral axis to section under consideration

          E_i = modulus of elasticity of lamina at that section

          EI = stiffness of whole section

Therefore stress in any particular lamina due to bending is given by:

$$\sigma = \frac{M}{Z} = \frac{M E_i x}{EI}$$

where the stress is dependent upon the modulus of elasticity of the lamina and its position from the neutral axis.

The shear stress is given by

$$\tau_s = \frac{Q \, \Sigma E_i A_i x}{EIb}$$

where the quantity $\Sigma E_i A_i x$ is the area stiffness moment of all laminae above section under consideration.

Q = vertical shear force at cross section being considered
b = width of beam at section

In this particular design it is convenient to transform the composite material consisting of different laminae into an equivalent section of one lamina and the cloth and chopped strand mat laminae will be transformed to an equivalent woven roving lamina (the width of the woven roving lamina is taken at 10 mm).

Area of woven roving lamina = $10 \times 4 = 40.0 \text{ mm}^2$

Area of chopped strand mat lamina   = $10 \times 1.5 \times \dfrac{7.0}{10.6} = 9.9056 \text{ mm}^2$

Area of cloth lamina   = $10 \times 1.0 \times \dfrac{11.0}{10.6} = 10.3774 \text{ mm}^2$

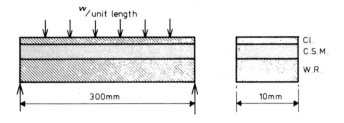

Figure 9.8.2

Second moment of area of woven roving lamina about its own centroidal axis = 53.3333 mm⁴.

Second moment of area of chopped strand mat lamina about its own centroidal axis = 1.8573 mm⁴.

Second moment of area of cloth lamina about its own centroidal axis = 0.8648 mm⁴.

The position of the neutral axis of the transformed area is obtained from:

$$\bar{x} = \frac{\Sigma A_i x_i}{\epsilon A_i}$$

$$= \frac{(40.0 \times 2) + (9.9056 \times 4.75) + (10.3774 \times 6)}{40 + 9.9056 + 10.3774}$$

$$= 3.14046 \text{ mm}$$

$$\text{Moment of inertia } I_{eq} = \Sigma \left[\frac{b_i t_i^3}{12} + A_i x_i^2 \right]$$

$$\begin{aligned}
&= (53.3333) + (40 \times 1.14046^2) + (1.8573) \\
&\quad + (9.906 \times 1.60954^2) + (10.3774 \times 2.85954^2) \\
&\quad + (0.8648)
\end{aligned}$$

$$= 218.5964 \text{ mm}^4$$

Section moduli to the various laminae are:

$$Z_i = \frac{I_{eq}}{y_i} \frac{E_{wr}}{E_i}$$

$$Z_{cl} = \frac{218.5964}{3.35954} \times \frac{10.6}{11.0} = 62.702 \text{ mm}^3$$

$$Z_{csm} = \frac{218.5964}{2.35954} \times \frac{10.6}{7.0} = 140.289 \text{ mm}^3$$

$$Z_{wr} = \frac{218.5964}{3.14046} \qquad = 69.606 \text{ mm}^3$$

The ultimate moments at which the stresses in each of the three laminae will reach their ultimate value are

$$M_{cl} = 216.0 \times 62.702 = 13543.416 \text{ N-mm}$$

$$M_{csm} = 140.0 \times 140.289 = 19640.460 \text{ N-mm}$$

$$M_{wr} = 195.0 \times 69.606 = 13573.170 \text{ N-mm}$$

Therefore the ultimate moment of resistance for the simply supported composite material beam is 13543.416 N-mm and the maximum uniformly distributed load for 10 mm width

$$w = \frac{M \times 8}{L^2}$$

$$= \frac{13543.416 \times 8}{300^2} = 1.204 \text{ N/mm}$$

The vertical shear force that the composite section is able to support is given by

$$Q = \frac{\tau_s \, I \, b \, E}{\Sigma E_i A_i x_i}$$

It is necessary to consider the shear force at the neutral axis and at the interface planes between the woven roving and chopped strand mat laminae and between the chopped strand mat and cloth laminae and then to calculate the uniformly distributed load on the minimum value.

(a) The vertical shear strength at the neutral axis within the woven roving laminae

$$Q = \frac{20.0 \times 218.5964 \times 10}{10 \times 3.14046 \times 1.57023} \times \frac{10.6}{10.6} = 886.537 \text{ N}$$

(b) Shear plane between the woven roving and chopped strand mat laminae

At the woven roving lamina surface

$$Q = \frac{20.0 \times 218.5964 \times 10}{(1.0 \times 10.3774 \times 2.85954) + (6.6038 \times 1.5 \times 1.60954)} \times \frac{10.6}{10.6}$$

$$= 958.375 \text{ N}$$

o

At the chopped strand mat laminae surface

$$Q = \frac{22.0 \times 218.5964 \times 6.6038}{45.6178} \times \frac{10.6}{7.0}$$

$$= 1054.2241 \text{ N}$$

(c) Shear plane between chopped strand mat and cloth laminae
At the chopped strand mat lamina surface

$$Q = \frac{22.0 \times 218.5964 \times 6.6038}{(10.3774 \times 1.0 \times 2.8595)} \times \frac{10.6}{7.0}$$

$$= 1620.626 \text{ N}$$

At the cloth laminae surface

$$Q = \frac{19.0 \times 218.5964 \times 10.377}{29.6747} \times \frac{10.6}{11.0}$$

$$= 1399.63 \text{ N}$$

Therefore the minimum ultimate interface shear strength (condition 'a')

$$= 886.537 \text{ N, say } 887 \text{ N}$$

and the ultimate uniformly distributed load/10 mm width on beam

$$w = \frac{2 \times Q}{L}$$

$$= \frac{2 \times 887}{300} = 5.91 \text{ N/mm}$$

The ultimate strength in bending controls the ultimate load bearing capacity of the composite material, therefore w equals 1.204 N/mm.

The following three examples deal with the design approach for sandwich beam construction. They illustrate:

(a) a thin face sandwich construction;
(b) a thick face sandwich construction;
(c) a sandwich construction in which the bending stiffness of the faces is high.

The behaviour of a loaded sandwich beam depends upon two non-dimensional parameters D_f/D and $D/L^2 D_Q$ and if these are significant, equation (6.15) must be used in preference to equation (6.12).

The following three examples illustrate the method of solution for problems related to beams which are simply supported and carry point loads at the centre. The first problem (example 9.9) concerns a thin face sandwich beam and it will be noted that the total deflection is given by the sum of the primary and secondary deformations without applying coefficients to the secondary deformations (i.e. equation (6.12) is used); this is due mainly to the span being relatively large for the size of beam (the value of $D/L^2D_Q = 0.0077$ and is small). The majority of the sandwich beams with G.R.P. faces and foamed plastics core come under this category.

The second problem (example 9.10) relates to a thick face sandwich beam made from G.R.P. faces and a foam core, and although there is significant stiffness in bending about the faces (the value of $D_f/D = 0.0357$) the value D/L^2D_Q is still small because the overall stiffness of the beam is low in value and the span is relatively large for the beam; the total deflection is given by equation (6.12).

The third example (example 9.11) has stiffer faces than the two previous ones; it has a smaller span and therefore the value of D/L^2D_Q is comparatively large in value; consequently the total deflection of the beam is given by equation (6.15).

Example 9.9

Figure 9.9.1 shows the cross section of a sandwich beam, Both faces are a composite of chopped strand glass fibre mat reinforcement and polyester resin (42% glass content by weight) and the core material is a polyurethane foam. The beam is simply supported with a span of 3600 mm and an overhang of 200 mm at each end. The central point load has a value of 2.25 kN. The maximum central deflection must not exceed 25 mm. The weight of the beam is to be ignored. The material properties are as follows:

E_f = 10.00 GN/m²   The width of the beam and the thickness of the
ν_f = 0.2   faces are assumed to be 640 mm and 6 mm
G_c = 20 MN/m²   respectively.

It is required to design the beam.

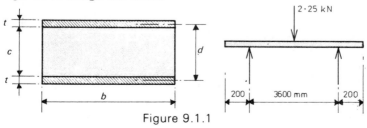

Figure 9.1.1

It will be assumed that the construction is a thin face sandwich beam, and at the end of the example this assumption will be justified.

In a sandwich beam the deflection is usually the criterion for design and as a first approximation the deflection due to bending moment is:

$$w_1 = \frac{W L^3}{48 EI}$$

$$EI = \frac{W L^3}{48 w_1} = \frac{2.25 \times 3600^3}{48 \times 25 \times 10^6}$$

$$= 87.48 \text{ kN m}^2$$

For a thin face sandwich beam

$$D = E_f \frac{bd^2 t}{2}$$

$$\therefore d = \sqrt{\left(\frac{87.48 \times 2 \times 10^6}{10.42 \times 640 \times 6} \right)}$$

$$= 66.126 \text{ mm}$$

As deflection due to shear has also to be considered assume

$$d = 70 \text{ mm}$$

It is now necessary to determine the properties of the section

$$I_f = \frac{bt^3}{6} = \frac{640 \times 6^3}{6} = 23040.0 \text{ mm}^4$$

$$\frac{bd^2}{c} = \frac{640 \times 70^2}{64} = 49000 \text{ mm}^2$$

$$I_{\text{total}} = \frac{btd^2}{2} = \frac{640 \times 6 \times 70^2}{2} = 9408000 \text{ mm}^4$$

$$\frac{E_f}{1-\nu^2} = \frac{10.00}{(1-(0.2)^2)} = 10.42 \text{ GN/m}^2$$

$$D = I_{\text{total}} \times \frac{E_f}{1-\nu^2} = \frac{9408000 \times 10.42 \times 10^3}{10^6} = 98031.35 \text{ MN-mm}^2$$

$$D_Q = G \times \frac{bd^2}{c} = \frac{20 \times 49000}{10^6} = 0.98 \text{ MN}$$

$$D_f = E_f I_f = \frac{10.42 \times 10^3}{10^6} \times 23040.0 = 240.0768 \text{ MN-mm}^2$$

$$\frac{D_f}{D} = \frac{240.08}{98031.35} = 0.00245$$

$$\left(\frac{1-D_f}{D}\right) = 0.99755$$

$$\left(\frac{1-D_f}{D}\right)^2 = 0.9951$$

As D_f/D is small compared with unity and

$$\frac{D}{L^2 D_Q} \text{ is very small} \left(\frac{D}{L^2 D_Q} = \frac{98031.35}{3600^2 \times 0.98} = 0.0077\right)$$

S_1 in equation (6.14) is close to unity and the standard formula for deflection (equation (6.11)) is used for a thin face sandwich beam.

The total deflection = the primary deformation + the secondary deformation.

$$= \frac{WL^3}{48EI} + \frac{WL}{4AG}$$

$$= \left(\frac{2.25 \times 3600^3}{48 \times 98031.35 \times 10^3}\right) + \left(\frac{2.25 \times 3600 \times 10^6}{4 \times 640 \times 64 \times 20 \times 10^3}\right)$$

$$= 22.309 + 2.472$$

$$= 24.781 \text{ mm}$$

The bending stress for a thin face sandwich beam is:

$$\sigma = \frac{M}{btd} = \frac{2.25 \times 3600}{4 \times 640 \times 6 \times 70 \times 10^{-6}}$$

$$= 7533.48 \text{ kN/m}^2$$

The shear stress in the core material is:

$$\tau = \frac{Q}{bd} = \frac{2.25 \times 10^6}{2 \times 640 \times 70.0}$$

$$= 25.1 \text{ kN/m}^2$$

It is necessary to justify the use of the simplified moment stiffness and shear stress formula.

The moment stiffness $D = E_f \dfrac{bd^2 t}{2}$ if:

(a) $100 > \dfrac{d}{t} > 5.77$

(b) $$\frac{E_f}{E_c}\frac{t}{c}\left(\frac{d}{c}\right)^2 > \frac{100}{6}$$

Both these conditions hold.

The shear-stress in the core material $= \dfrac{Q}{bd}$ if:

(a) $$\frac{E_f}{E_c}\frac{t}{c}\frac{d}{c} > \frac{100}{4};$$

(b)   The moment stiffness above may be used.

Both these conditions hold.

Example 9.10

A sandwich beam of similar manufacture to that in example 9.9 and with dimensions shown in Figure 9.10.1 (where $d/t < 5.77$) is simply supported over a span of 3600 mm. It is to support a central point load of 0.7 kN. The overhang is 200 mm at each end. The maximum central deflection must not exceed 25 mm. The weight of the beam is to be ignored. The material properties are as follows:

$$E_f = 10.00 \text{ GN/m}^2$$

$$v_f = 0.2$$

$$G_c = 20 \text{ MN/m}^2$$

It is required to design the beam.

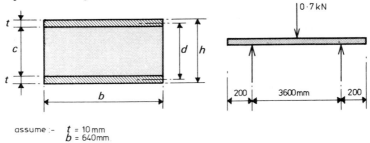

assume :- t = 10 mm
            b = 640mm

Figure 9.10.1

The solution of this problem is similar to that of example 9.9 and only the main results will be given.

The construction is a thick face sandwich beam.

The first approximation, to enable the sizes of the beam to be estimated, is to ignore deflection due to shear

$$w_1 = \frac{WL^3}{48 EI}$$

$$\therefore D = \frac{0.7 \times 3600^3}{48 \times 25 \times 10^6}$$

$$= 27.216 \text{ kN m}^2$$

$$D = E_f I_{total}$$

$$I_{total} = \frac{D}{E_f}$$

$$\frac{bt^3}{6} + \frac{btd^2}{2} = \frac{D}{E_f}$$

$$d = \sqrt{\left[\left(\frac{27.216 \times 10^6}{10.42} - \frac{640 \times 10^3}{6}\right)\frac{2}{640 \times 10}\right]}$$

$$= 27.980 \text{ mm}$$

Assume $d = 30$ mm (to allow for deflection due to shear)

$$I_f = \frac{bt^3}{6} = \frac{640 \times 10^3}{6} = 106666.7 \text{ mm}^4$$

$$\frac{bd^2}{c} = \frac{640 \times 30^2}{20} = 28800 \text{ mm}^2$$

$$I_{total} = \frac{bt^3}{6} + \frac{btd^2}{2}$$

$$= 2986666.7 \text{ mm}^4$$

$$\frac{E_f}{1-\nu^2} = \frac{10.00}{1-(0.2)^2} = 10.42 \text{ GN/m}^2$$

$$D = I_{total} \times \frac{E_f}{1-\nu^2} = 31121.065 \text{ MN mm}^2$$

$$D_Q = \frac{G bd^2}{c} = 0.576 \text{ MN}$$

$$D_f = E_f I_f = 1111.460 \text{ MN mm}^2$$

$$\frac{D_f}{D} = \frac{1111.460}{31121.065} = 0.0357$$

$$\frac{D}{L^2 D_Q} = \frac{31\ 121.065}{3600^2 \times 0.576} = 0.00417$$

$\frac{D_f}{D}$ is small compared with unity and $\frac{D}{L^2 D_Q}$ is very small.

Therefore S_1 in equation (6.15) is close to unity and the standard formula for deflection (equation (6.12)) is used for the thick face sandwich beam.

The total deflection = the primary deflection + the secondary deflection.

$$= \frac{WL^3}{48EI} + \frac{WL}{4AG}$$

$$= \left(\frac{0.7 \times 3600^3}{48 \times 31121.065 \times 10^3}\right) + \left(\frac{0.7 \times 3600 \times 10^6}{4 \times 20 \times 640 \times 20 \times 10^3}\right)$$

$$= 21.817 + 2.461$$

$$= 24.278 \text{ mm}$$

Example 9.11

Figure 9.11.1 shows the cross section of a sandwich beam. Both faces are made of sheet steel; the top being a trough section of 1.22 mm thickness and the bottom a sheet section of 1.5 mm thickness. The core material is 28 mm thickness and is a low density foamed polyurethane. The width of the beam is 1000 mm and the beam is simply supported with a span of 1200 mm and an overhang of 200 mm at each end. The central point load has a value of 40 kN. Ignore the weight of the beam. The material properties are as follows.

$E_f = 205 \text{ GN/m}^2$

$v_f = 0.3$

$G_c = 20 \text{ MN/m}^2$

It is required to design the beam.

The properties of the section are:

$$\bar{x} = \frac{\Sigma Ax}{\Sigma A}$$

Length of trough section $= (30 \times 2 + 2 \times 20 \times \sqrt{2})$

$$= 116.569 \text{ mm}$$

Weight of core material $= 75.0 \text{ kg/m}^3$

Weight of steel $= 7700 \text{ kg/m}^3$

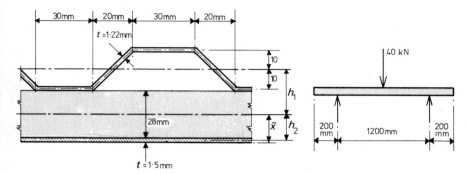

Figure 9.11.1

$$\therefore \bar{x} = \frac{116.569 \times 1.22 \times 7700 \times 39.36 + 75 \times 28 \times 100 \times 14.75}{116.569 \times 1.22 \times 7700 + 75 \times 28 \times 100 + 100 \times 1.5 \times 7700}$$

$$= 18.7796 \text{ mm}$$

$$h_1 = 20.58 \text{ mm; and } h_2 = 18.78 \text{ mm (both by calculation)}$$

$$c = 28 \text{ mm}; \quad d = h_1 + h_2 = 39.36 \text{ mm}$$

$$\frac{bd^2}{c} = \frac{1000 \times 39.36^2}{28} = 55\,328.914 \text{ mm}^2$$

$$I_{f1} = \left[(30 \times 1.22 \times 10^2 \times 2) + \left(\frac{1.22 \times 1.414 \times 20^3 \times 2}{12} \right) \right] \times 10$$

$$= 96201.07 \text{ mm}^4 \text{ (for whole width of beam)}$$

$$I_{f2} = \frac{1000 \times 1.5^3}{12}$$

$$= 281.25 \text{ mm}^4 \text{ (for whole width of beam)}$$

$$I_f = 96\,201.07 + 281.25$$

$$= 96\,482.32 \text{ mm}^4 \text{ (for whole width of beam)}$$

$$\begin{aligned} I \text{ total} &= (96\,482.32) + [116.569 \times 1.22 \times (20.58042)^2 \times 10] + \\ &\quad [100 \times 1.5 \times (18.7796)^2 \times 10] \end{aligned}$$

$$= 1\,227\,844.6 \text{ mm}^4 \text{ (for whole width of beam)}$$

$$E_f/(1 - v^2{}_f) = 225\,274.73 \text{ MN/m}^2$$

$$D_Q = 20 \times 55\,328.914 \times 10^{-6} = 1.107 \text{ MN}$$

$$D_f = 225\,274.73 \times 96\,482.32 \times 10^{-6} = 21735.029 \text{ MN mm}^2$$

$$D = 225\ 274.73 \times 1\ 227\ 844.6 \times 10^{-6} = 276\ 602.36\ \text{MN mm}^2$$

$$D_f/D = \frac{21\ 735.029}{276\ 602.36} = 0.07858$$

$$(1-D_f/D) = 0.921$$

$$(1-D_f/D)^2 = 0.8482$$

$$D/L^2 D_Q = \frac{276\ 602.36}{1200^2 \times 1.107} = 0.17352$$

Overhang $L_1 = 200\ \text{mm}$

$$\theta = \tfrac{1}{2}\,[(D/L^2 D_Q)\,(D_f/D)\,(1-D_f/D)]^{-\frac{1}{2}}\ \text{(from Appendix 6.1)}$$

$$= \tfrac{1}{2}\,[0.17352 \times 0.07858 \times 0.921]^{-\frac{1}{2}}$$

$$= 4.461793$$

$$\phi = \frac{4.461793 \times 2 \times 200}{1200} = 1.487$$

$$\beta_1 = \frac{\sinh\theta - (1-\cosh\theta)\tanh\phi}{\sinh\theta\,\tanh\phi + \cosh\theta}\ \text{(from Appendix 6.1)}$$

$$= \frac{43.374123 - (1-43.385679) \times 0.9027685}{43.374123 \times 0.9027685 + 43.385679}$$

$$= 0.9890$$

$$S_1 = 1 - \frac{\sinh\theta + \beta_1\,(1-\cosh\theta)}{\theta}$$

$$= 1 - \frac{43.374123 + 0.9890 \times (1-43.385679)}{4.461793}$$

$$= 0.674$$

From equation (6.15) the total central deflection

$$= \frac{WL^3}{48D} + \frac{WL}{4D_Q}\left(1-\frac{D_f}{D}\right)^2 S_1$$

$$= \left(\frac{40 \times 1200^3}{48 \times 276602.36 \times 10^3}\right) + \left(\frac{40 \times 1200 \times 0.8482 \times 0.674}{4 \times 1.107 \times 10^3}\right)$$

$$= 5.2060 + 6.1972$$

$$= 11.4032\ \text{mm}$$

The equivalent equation for the total central deflection of a beam carrying a uniformly distributed load is given in Appendix 6.1, also given are the relevant coefficients.

Example 9.12

Determine the critical buckling load for a sandwich panel manufactured from faces consisting of composite of chopped strand mat reinforcement and polyester resin (30% glass content by weight) and core material of polyurethane foam. The panel is simply supported on four sides. The modulus of elasticity in compression of the material of the faces is 7.0 GN/m², Poisson's ratio is 0.2 and the modulus of rigidity of the core material is 20 MN/m². The length and width of the panel are 2m and 1m respectively.

Figure 9.12.1 shows the sandwich beam.

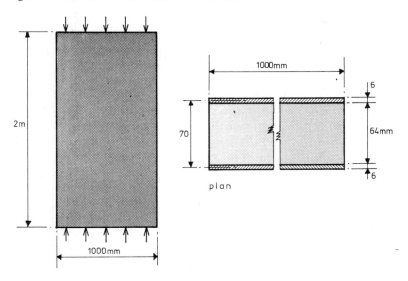

Figure 9.12.1.

The critical buckling load for a thin face sandwich panel

$$P_{cr} = \frac{\pi^2 D_2}{b^2} K \text{ (equation (6.23))}$$

where the buckling coefficient K is determined from Figure 6.13 in terms of the length/width ratio of the panel and the quantity ρ where ρ is:

$$\frac{\pi^2}{b^2} \left(\frac{E_f d^2 t}{2(1-v_f^2)} \right) \left(\frac{c}{G_c d^2} \right)$$

$$\text{Therefore } \rho \quad = \quad \frac{\pi^2}{1000^2}\left(\frac{7.0 \times 70^2 \times 6 \times 10^3}{2(1-0.2^2) \times 10^6}\right)\left(\frac{64 \times 10^6}{20 \times 70^2}\right)$$

$$= \quad 0.6909$$

$$D_2 = \frac{E_f t d^2}{2(1-v_f^2)} = \frac{7.0 \times 6 \times 70^2 \times 10^3}{2(1-0.2^2) \times 10^6} \quad \text{MN/mm}^2$$

$$= \quad 107.1875 \text{ MN/mm}^2$$

Buckling coefficient $K = 1.5$

$$\therefore P_{cr} \quad = \quad \frac{\pi^2 D_2}{b^2} \quad K$$

$$= \quad \frac{\pi^2 \times 107.1875}{10^6} \quad \times 1.5 \times 10^6 \text{ N} \quad = \quad 1586.96 \text{ N}$$

Figure 6.13 may only be used if D_f/D and $D/L^2 D_Q$ give a point in Figure 6.9 which is close to the line ABCG. It is, therefore, necessary to check this condition:

$$D_f = E_f I_f$$

where $\quad I_f = \dfrac{6^3}{6} \quad = \quad 36.0 \text{ mm}^4$

so $\quad D_f = \dfrac{7.0 \times 36.0 \times 10^3}{(1-0.2^2) \times 10^6} = 0.2625 \text{ MN mm}^2$

$$D = I_{\text{total}} \times \frac{E_f}{(1-v_f^2)}$$

$$I_{\text{total}} = \frac{td^2}{2} = \frac{6 \times 70^2}{2} = 14700.0 \text{ mm}^4$$

$$D = \frac{14700.0 \times 7.0 \times 10^3}{(1-0.2^2) \times 10^6} = 107.1875 \text{ MN mm}^2$$

$$\therefore D_f/D = 0.00245$$

$$D_Q = \frac{Gd^2}{c} = \frac{20 \times 70^2}{64 \times 10^6} \text{ MN} = 0.0015312 \text{ MN}$$

$$D/L^2 D_Q \quad = \quad \frac{107.1875}{0.0015312 \times 2000^2} = 0.0175$$

The above values of D_f/D and $D/L^2 D_Q$ give a point on the line ABCG of Figure 6.9.

9.2 Detailing Examples

The following examples of the detailing of G.R.P. components are taken from Civil Engineering projects which have been constructed in recent years.

Example 9.13

Mondial House, London
Client: Department of the Environment
Architects: Hubbard Ford and Partners, London, U.K.
G.R.P. Consultants: Peter Hodge and Associates, Crewkerne, Somerset,

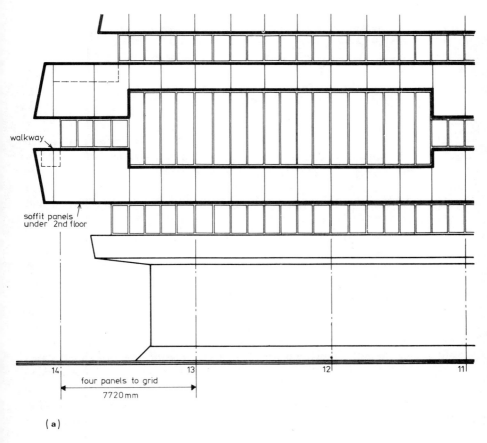

(a)

Figure 9.13.1(a) Mondial House; elevation; (Peter Hodge and Associates)

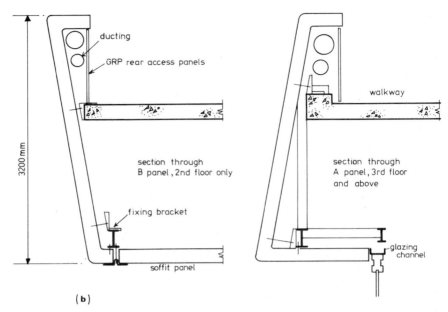

ducting

GRP rear access panels

walkway

3200 mm

section through
B panel, 2nd floor only

section through
A panel, 3rd floor
and above

fixing bracket

glazing
channel

soffit panel

(b)

Figure 9.13.1(b) Details

Figure 9.13.1 shows a part elevation of Mondial House, a section through B panel, 2nd floor, and a section through A panel, 3rd floor and above. Figures 9.13.2, 9.13.3 and 9.13.4 show typical section details. The outer skin comprises a gel coat using an isopthalic resin, pigmented white, with an ultraviolet stabilizer backed up with a laminate using chopped strand mat glass reinforcement at the rate of 900 g/ m² and a self extinguishing laminate resin at the rate of 2700 g/ m².

Thermal insulation and rigidity are obtained by using a core material of rigid polyurethane foam, bonded to the outer skin and covered on the back with a further glass reinforced laminate. Further rigidity is obtained by the use of lightweight top hat section beams, manufactured as thin formers and incorporated and overlaminated into the moulding as manufacture proceeds.

The face of the panel is reeded on the vertical surfaces in order to mask any minor undulations, to provide channels for the water to run off, thereby enhancing the self cleansing properties and to give the effect of a matt panel without reducing the high surface finish necessary to maintain the whiteness of the panels.

The joints between the panels take into account three factors, these are: the thermal movement between the panels, the fixing tolerances and the prevention of ingress of air and water into the building through the joints.

The joint is basically a mechanical one and the mean gap between the panels

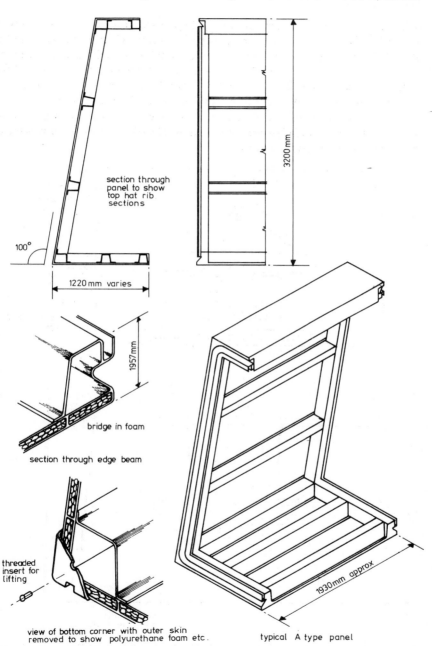

section through
panel to show
top hat rib
sections

100°

1220 mm varies

3200 mm

1957 mm

bridge in foam

section through edge beam

threaded
insert for
lifting

view of bottom corner with outer skin
removed to show polyurethane foam etc.

typical A type panel

1930 mm approx

Figure 9.13.2 Details (Peter Hodge and Associates)

is 32 mm, the minimum being 26 mm and the maximum 38 mm; this allows a tolerance of 12 mm for fixing. In the vertical joint is a U-shaped drain channel of G.R.P. attached to a back baffle; this channel bridges the gap between the panels and holds captive two Neoprene cord seals against the rear face of the panel edge beams. The initial force of wind and rain is broken by a loose wide Neoprene baffle which is fitted in front of the U-shaped drain channel.

The panels are fixed to the building at slab level in two places and to the rolled steel joists suspended beneath the slab. The panels are also supported at the soffit by the rolled steel joist.

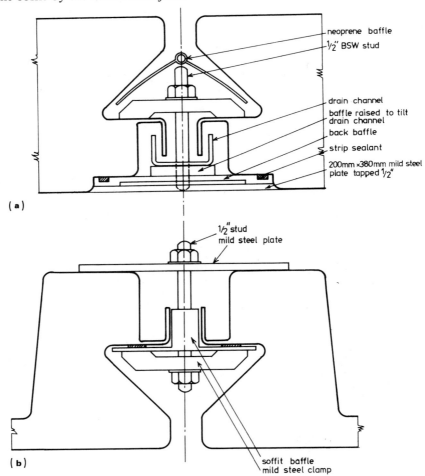

Figure 9.13.3 (a) Detail at joint between two panels: top return; (b) section through soffit joint (Peter Hodge and Associates)

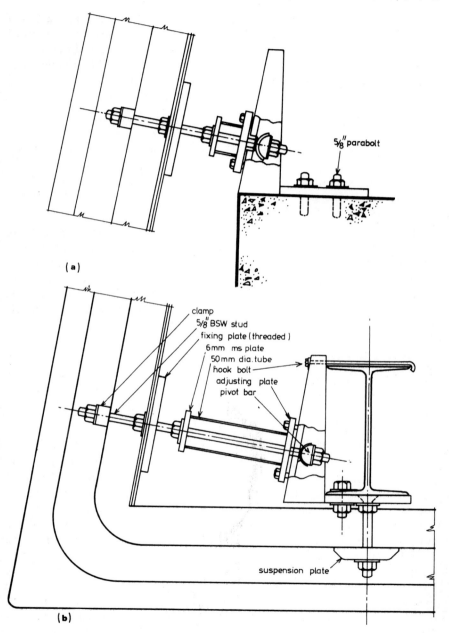

(a)

clamp
5/8" BSW stud
fixing plate (threaded)
6mm ms plate
50mm dia. tube
hook bolt
adjusting plate
pivot bar

5/8" parabolt

suspension plate

(b)

Figure 9.13.4 (a) Typical A panel fixing at slab; (b) fixings to I beam (Peter Hodge and Associates)

P

The method of fixing is achieved by clamps which impinge on the edges of the panels. The clamps are fitted into specially formed grooves and are fixed to the main bracket by a single 15 mm diameter stud. The main bracket is fixed to the slab by 'parabolts' and to the R.S.J. by means of 12 mm bolts. The bracket assembly allows for a considerable amount of movement and adjustment, both laterally and at right angles to the building.

Example 9.14

American Express European Headquarters Building, Brighton, U.K.
Clients: American Express Realty Management Co. Ltd.
Architects: Gollins, Melvin Ward Partnership, London, U.K.
Design and Specification Consultants: Peter Hodge and Associates, Crewkerne, Somerset, U.K.

Figure 9.14.1 shows a section through the G.R.P. load bearing cladding unit. An interesting aspect of the design of this building is the use of G.R.P. in several small components which were required for each panel fixing. An example is the clamps which hold the panels to the columns and which are manufactured from 4200 g/m² laminates in the form of 'U' section channels.

G.R.P. was also employed for decorative column cladding, soffit panels and for ancillary items such as drinking fountains.

Example 9.15

Covent Garden Flower Market, London, U.K.
Clients: Covent Garden Market Authority
Architects: Gollins Melvin Ward Partnership
Consulting Engineers: Clarke Nicholls & Marcel
Specialist Consultants on G.R.P.: Nachshen Crofts and Leggatt

The construction of the flower market is a steel space frame on a diagonal grid of 3.5 m with an external cantilevered canopy integrated with the roof steelwork. The roof is clad with double skin G.R.P. units with translucent panels in the base of each. Services are run within the roof framework.

Figure 9.15.1 shows the general roof construction with a detail of the connection of the G.R.P. roof panels and the skeletal space frame. The manufacture and fabrication of the G.R.P. roof panels was by the S.A.S. process (Suction and Squeeze), which was specifically developed for the Covent Garden project. This process has been described by Roach ('The manufacturer's view of the general use of plastics panels as structural and non-load bearing units', Chapter 3, *The Use of Plastics for Load Bearing and Infil Panels,* (ed.) Hollaway, L. Manning Rapley Publishing, 1976).

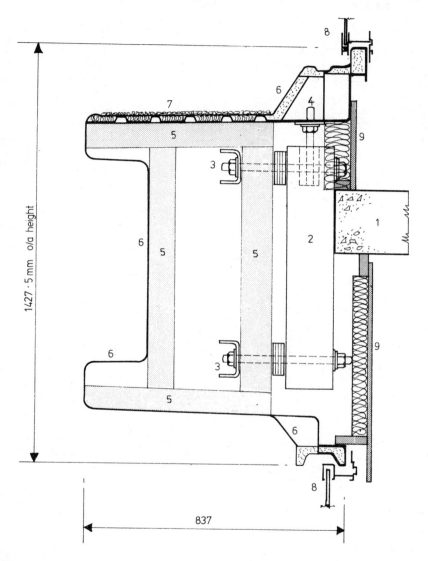

Figure 9.14.1 American Express Building; section through G.R.P. load bearing cladding panel/beam. 1. R.C. slab; 2. R.C. corbel; 3. restraint clamp fixing; 4. load bearing jacking stud; 5. box and top hat section integral G.R.P. framework at intervals throughout length of beam; 6. continuous outer G.R.P. skin construction/profile; 7. calcine flint covered window cleaning walkway; 8. curtain walling (detail) supported by the G.R.P.; 9. firestop and insulation

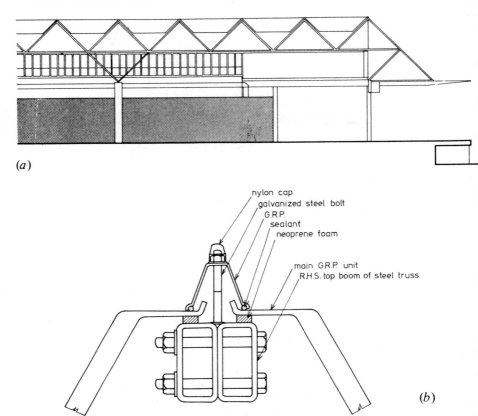

(a)

nylon cap
galvanized steel bolt
G.R.P.
sealant
neoprene foam

main G.R.P. unit
R.H.S. top boom of steel truss

(b)

Figure 9.15.1 Details of Covent Garden Flower Market; (a) section of structure; (b) key connection of G.R.P. roof panels and the steel space frame

Appendix

Definitions

The use and range of applications of fibre reinforced matrix materials have considerably increased over recent years. Consequently, a new technology has been established involving a number of specialized processes and techniques. Since those working in the area may have widely differing backgrounds, it is useful to define the more important terms used in this book.

Matrix   A bonding material which adheres to and contains the fibres. Many materials such as the thermoplastic and thermosetting resins, metals, glass or ceramic materials, can form a matrix. Resins are the most widely used.

Thermoplastic plastics   A material is thermoplastic when it can be softened by heating and hardened by cooling without undergoing a chemical change.

Thermosetting plastics   A material is thermosetting when it can be changed or has been changed into a hard infusible product by a non-reversible chemical reaction initiated by the use of heat or curing agents.

Monomer   The monomer is the low molecular weight starting material from which the polymer is formed.

Addition polymers   Polymers formed into long chain molecules by the chemical reaction of one or more types of monomer units, each of which has a double bond prior to polymerization.

Condensation polymers   Polymers formed by the chemical reaction of at least two monomer units with, in most cases, production of a by-product of low molecular weight.

Co-polymer   An addition polymer of at least two monomers.

Polymerization   Polymerization is the chemical reaction involved when, say, a liquid polyester resin sets to a solid. A comparatively simple chain molecule becomes a highly complex three-dimensional one.

Degree of polymerization   An impression of the length of the average molecular chain in a polymer; assessed by an estimate of the average molecular weight and usually stated in terms of the number of repeating units in a chain.

217

Hydrophilic   The property of possessing strong affinity for water.

Glass transition temperature   The temperature at which a sudden change in slope of various physical properties versus temperature curves occur (commonly measured in terms of the standard heat distortion temperature). It very nearly approximates the temperature below which a polymer fails in a brittle manner and above which it behaves as a leathery or rubbery solid.

Elastomer   An elastomer is any member of a class of synthetic polymeric substances possessing rubber like qualities (especially the ability to regain shape after deformation), toughness and having a glass transition temperature well below ambient temperature.

Plasticizers   Materials deliberately added to polymers to reduce their stiffness.

Fibre   Any material in an elongated form such that the ratio of its minimum length to its maximum average transverse dimension is 10:1, its maximum cross-sectional area is 1.975×10^{-3}mm² (corresponding to a circular cross-section of 0.25 mm diameter) and its transverse dimension is not greater than 0.25 mm.

Fibre composite material   A material consisting of two or more distinct physical phases, one of which is a fibrous phase dispersed in a continuous matrix phase.

Filament   A continuous fibre.

Whisker   Any material that fits the definition of a fibre and is a single crystal.

Wire   A metallic filament.

Crystallite   The most rudimentary form of an embryonic crystal that can be identified as a certain species under the microscope.

PAN   A precursor used in the manufacture of carbon fibres. PAN is the abbreviation for polyacrylonitrile.

Yarn or tow   A number of filaments in a bundle which can be handled as a single unit. A tow is usually bigger than a yarn, having thousands of filaments, whereas a yarn usually has a few hundred filaments. A yarn may be spun and twined from staple fibre, but a tow is formed from constant filaments.

Pre-preg (pre-impregnated fibre)—An intermediate product consisting of fibres or tows which have been coated with a matrix material such as resin. The fibres are aligned in the majority of cases to give a flat sheet or tape. Usually the resin is not fully cured so that the aggregate remains flexible and the sheet can be built up in piles to form a composite.

Strand   This is associated with filaments of glass fibre. A strand is a bundle of 204 filaments of glass fibre. The diameter of a filament is up to 1/400 mm.

Chopped strands   These are made from continuous strands which are chopped into short lengths (usually 50 mm).

Chopped strand mats   These are chopped strands which are held together by means of a size. The strands are completely randomly orientated and the mats are of uniform thickness. They are produced in various sizes, the most usual being 1 oz/ft² and 1½ oz/ft² (300 g/m² to 400 g/m²).

Woven rovings   These are continuous strands which may be unidirectionally or bi-directional orientated.

Woven cloth   This is a more refined product than above. It is usually a bi-directional reinforcement.

Continuous fibre reinforcement   Continuous fibres may be defined as fibres which are continuous throughout the whole length of the laminate, resulting in the load being applied directly to them; the stress throughout the length of the fibre is constant.

Discontinuous fibre reinforced plastics   Discontinuous fibre reinforced plastics refer to plastics whose reinforcing fibres have length-to-diameter (l/d) ratios (known as aspect ratios) varying between 100 and 5000. Whiskers reinforcement have l/d ratios between approximately 150 and 2500. Discontinuous glass fibre reinforced plastics which consist of premix thermosetting resins, chopped strand mat-reinforced thermoset resins, and fibre-reinforced thermoplastics resins have fibres with l/d ratios between 150 and 5000. The ultimate strength and modulus of short-fibre-reinforced composites can approach the values for continuous fibre composites providing that the short filaments can be aligned unidirectionally and that their length is much greater than the critical length (l_c) required for shear stress transfer.

Critical length   The critical length of a fibre is the length which is required for the fibre stress to develop its maximum value when under a particular load condition.

Sandwich beam   A sandwich beam has two faces, each of thickness t separated by a layer, or core, of low density material of thickness c.

Very thin face   (associated with sandwich beams)   One in which the stiffness in bending about its own axis is taken as zero and is sufficiently thin to assume d equal to c.

Thin face   (associated with sandwich beams)   One in which the stiffness in bending about its own axis is assumed to be zero, but the thickness of the faces is assumed to have a finite value so that d is not equal to c.

Thick face   (associated with sandwich beams)   One in which the stiffness in bending about its own axis is significant and the thickness of the faces is assumed to have a finite value so that d is not equal to c.

Antiplane core   (associated with sandwich beams) one in which $\sigma_{xx} = \sigma_{yy} = \sigma_{xy} = 0$. Consequently the shear stresses σ_{zx} and σ_{yz} are independent of z.

Flexural rigidity   (associated with sandwich beams or struts)   neglecting the local bending stiffness of the faces is:

$$D_1 = E_f \, b \, t \, d^2 / 2$$

Flexural rigidity   (associated with sandwich panels) with cylindrical bending and neglecting the local bending stiffness of the faces is:

$$D_2 = E_f \, t \, d^2 / 2 \, (1 - \nu_f^2)$$

Isotropic material   This term indicates that the material properties at a point in the body are not a function of orientation. The material properties remain constant regardless of the reference co-ordinate system at a point. As a result the material properties are constant in any plane which passes through a point in the material. All planes which pass through a point in the isotropic material are planes of material property symmetry. Two independent elastic constants are necessary to write the Hooke's Law relationship for two-or three-dimensional stress states.

Orthotropic material   Only three mutually perpendicular planes of material property symmetry may be passed through a point. Four independent elastic constants must be determined.

Anisotropic material   There are no planes of material property symmetry which pass through a point. Therefore, the material constants at a point change as the co-ordinate system is rotated at the point. An anisotropic material in a plane stress state has six independent elastic constants.

Isotropic and orthotropic materials are specialised cases of material which have a higher degree of symmetry than anisotropic materials.

Homogeneous material   The material properties do not change from point to point in the body, although they may change with a co-ordinate rotation at the points, thus as long as the material properties are the same for the same co-ordinate position the material is homogeneous. If this is not true the material is said to be *heterogeneous.*

Yield point   The maximum stress recorded in a tensile or compressive test of a ductile specimen prior to entering the inelastic region of the material.

Yield stress   A term denoting the yield strength or yield point of a material as defined above.

Tangent modulus   The slope of the stress-strain curve of a material in the inelastic region, at any given stress level as determined by the compression test of a small specimen.

Buckle   The process of wrinkling or bulging of a member as a result of elastic or inelastic strain.

Buckling load   The load at which a compression member will collapse in service or fails in a loading test.

Critical load   The load at which bifurcation occurs. It is determined by theoretical stability analysis.

Bifurcation A term related to the load-deflection relationship of a straight, axially loaded strut at critical load. It is a point at which divergent equilibrium states become possible and results at a branch point in the plot of axial load against lateral deflection from which two alternative load deflection plots are theoretically valid.

Effective width This term refers to the reduced width of plate or angle which, assuming uniform stress distribution, will give the same structural behaviour as the actual section of the plate or angle and the actual non-uniform stress distribution.

Index